Praise f

"His writings are infl… in many parts of the world...."

—*El Mercurio*, Santiago

"This book should provoke debate in the circles of the Left, promoting fresh dialogue among those searching for an understanding with which to confront the present advance of neoliberalism and to open paths toward change."

—*Revista Punto Final*, Santiago

Praise for Silo's *Tales for Heart and Mind*

Winner of a 1994 Benjamin Franklin Book Award

"...a large, liberally illustrated volume...Silo spins dreamy fantasies...charming, healing mind-trips that speak meaningfully to our daily predicaments...Berry's black-and-white illustrations convey childlike wonder and a sense of mystery."

—*Publishers Weekly*

"...a unique contribution to the mix of history, fable, racial memories and contemporary experience. His passages – short stories, scenes, prose poems – tantalize the mind and generate understandings."

—James A. Michener

"The great age of illustrated adult literature is long past, but unexpected pleasures issue from occasional books whose drawings illuminate the prose and whet a reader's fascination. Such a work is *Tales for Heart and Mind*...Silo is part Shakespeare, part Dante, part Lewis Carroll, but also entirely himself."

—*Fort Worth Star-Telegram*

"...unusual stories gleaming like gems in the crown of the imagination of Argentine writer Silo...*Tales for Heart and Mind* is more than entertainment, it is food for the soul."

—*The Palm Beach Post*

"Adults who have forgotten how to appreciate a good story – especially one in which they can imagine themselves participating – would do well to pick up *Tales for Heart and Mind*, playful wisdom by Silo..."

—*The Columbus Dispatch*

"Silo's *Tales for Heart and Mind* is a wonderful guide which is hard to categorize: a collection of stories that blend philosophical reflections with fables and insights on contemporary experiences...."

—*The Bookwatch*

Letters to My Friends

Letters to My Friends

On Social and Personal Crisis in Today's World

Silo

Translated from the Spanish
by Paul Tooby

Latitude Press
San Diego

This book is part of the New Humanism Series of contemporary fiction and nonfiction addressing the growing interest in humanizing individual and social life in today's world. A free catalog, additional copies of this book, or other titles in the New Humanism Series are available from Latitude Press, P.O. Box 231516, Encinitas, CA 92023-1516. Write or call us at 800-LATITUDE (800-528-4883).

Latitude Press is committed to publishing books that reflect a contemporary humanist approach to the problems of today's world, and we apply the net income from these titles toward efforts to build a human and nonviolent society.

This edition contains the complete and unabridged text of this work. Nothing has been omitted.

LIBRARY OF CONGRESS CATALOGING-IN-PUBLICATION DATA
Silo, 1938–
[Cartas a mis amigos. English]
Letters to my friends: on social and personal crisis in today's world / Silo; translated from the Spanish by Paul Tooby.
1st ed.
p. cm.
Includes bibliographical references.
ISBN 1-878977-23-7 (pbk.: acid-free recycled paper)
1. Humanism. 2. History, Modern – 1945-
3. Social problems. 4. Ethical problems I. Title.
BL2747.6.S5513 1994
144 – dc20 94-20708 CIP

Published by Latitude Press.
Printed in the U.S.A. on acid-free recycled paper meeting the guidelines of the Committee on Production Guidelines for Book Longevity of the Council on Library Resources.

First Edition
A B C D E

Contents

Introduction ix

First Letter to My Friends 1

1. The Present Situation 3
2. The Alternative of a Better World 3
3. Social Evolution 4
4. Future Experiments 6
5. Change and Relationships Among People 7
6. A Tale for Aspiring Executives 8
7. Human Change 10

Second Letter to My Friends 15

1. Some Positions Regarding the Present Process of Change 16
2. Individualism, Social Fragmentation, and the Concentration of Power in a Few 18
3. Characteristics of the Crisis 20
4. Positive Factors of Change 23

Third Letter to My Friends 27

1. Change and Crisis 27
2. Disorientation 27
3. Crisis in the Life of Each Person 28
4. The Need to Give Direction to One's Life 28
5. Direction in Life and Changing One's Situation 28
6. Coherent Behavior 29
7. The Two Proposals: Coherence and Solidarity 30
8. Reaching All of Society Starting with One's Immediate Environment 31
9. The Social Environment in Which One Lives 32
10. Coherence as a Direction in Life 33

11. Proportion in One's Actions as a Step Toward Coherence 34
12. Well-Timed Actions as a Step Toward Coherence 34
13. Growing Adaptation as an Advance Toward Coherence 36

Fourth Letter to My Friends 39

1. The Starting Point for Our Ideas 41
2. The Human Being: Nature, Intention, and Opening 42
3. The Human Being: Social and Historical Opening 43
4. The Transforming Action of the Human Being 44
5. Overcoming Pain and Suffering as Basic Vital Projects 46
6. Image, Belief, Look, and Landscape 47
7. The Generations and Historical Moments 49
8. Violence, the State, and the Concentration of Power 51
9. The Human Process 52

Fifth Letter to My Friends 55

1. The Most Important Issue: To Know If One Wants to Live, and In What Conditions 56
2. Human Liberty: Source of All Meaning 57
3. Intention: Orientor of Action 57
4. What Should We Do With Our Lives? 58
5. Moral Consciousness and Short-Term Interests 61
6. Sacrificing One's Objectives for Circumstantial Success: Some Habitual Errors 62
7. The Kingdom of the Secondary 64

Sixth Letter to My Friends 67

Document of the Humanist Movement 69

I. Global Capital 69
II. Real Democracy Versus Formal Democracy 73
III. The Humanist Position 75
IV. From Naive Humanism to Conscious Humanism 77
V. The Anti-Humanist Camp 78
VI. Humanist Action Fronts 79

Seventh Letter to My Friends 81
1. Destructive Chaos or Revolution 82
2. Of What Revolution Are We Speaking? 85
3. Action Fronts in the Revolutionary Process 88
4. Revolutionary Process and Revolutionary Direction 92

Eighth Letter to My Friends 95
1. The Need to Redefine the Role of the Armed Forces 95
2. Continuing Factors of Aggression in This Period of Reduced Tensions 97
3. Internal Security and Military Restructuring 98
4. A Review of the Concepts of Sovereignty and Security 98
5. The Legality and Limits of Established Power 100
6. Military Responsibility to Political Power 101
7. Military Restructuring 103
8. The Military's Position in the Revolutionary Process 105
9. Considerations on the Military and Revolution 107

Ninth Letter to My Friends 109
1. Violations of Human Rights 109
2. Human Rights, Peace, and Humanitarianism as Pretexts for Intervention 112
3. The Other Human Rights 114
4. The Universality of Human Rights and the Cultural Thesis 117

Tenth Letter to My Friends 123
1. Destructuring and Its Limits 124
2. Some Important Areas of the Phenomenon of Destructuring 128
3. Targeted Action 134

Notes 137

Acknowledgements

In translating this collection of letters into English the assistance of numerous people was indispensable, in particular Daniel Zuckerbrot, without whose meticulous reading of the manuscript and extensive comments this work would not have been realized. Special thanks also go to Jorge van Schouwen for help in producing the first drafts of many of the letters and checking numerous translation issues and to Bill Pfender for his careful research.

Introduction

The letters collected here in book form were published separately as the author completed each one. No more than three years have elapsed from the time the author wrote the first of these letters in February of 1991 to when he completed the tenth and final one in December of 1993, but during this time sweeping transformations have taken place in almost every field of human endeavor. If the pace of change continues to increase as it did over this period, a reader from coming decades will have serious difficulty in understanding the world context to which the author continually makes reference, and will hence fail to grasp many of the ideas expressed in these writings.

I would recommend that these hypothetical readers of the future refer to a summary of the events that took place between the years 1991 and 1994, and that they seek an ample understanding of those times, of the economic and technological developments, the famines and conflicts, the trends in advertising and fashion. I would further suggest that they listen to the music of that period, that they view photographs of the architectural and urban landscapes, that they observe the overcrowding in the macro-cities, the vast human migrations, the deterioration of the environment, and the general way of life during that curious historical moment. Above all I would urge that they penetrate the fog of pronouncements and counter-pronouncements issued by the experts and formers of opinion: the philosophers, sociologists, and psychologists of that cruel and stupid age.

Although these letters address a *particular* present situation, they are unmistakably written with an eye toward the future, and

I believe that only from that future will it be possible to confirm or refute them.

The collection of letters in this work follows no general plan; it consists rather of a series of occasional expository writings that one may read in any order. Nevertheless, we might classify them as follows:

A. The first three letters focus on the experiences of the individual living in the midst of a global situation that is growing more complex day by day.
B. The fourth letter presents the general structure of ideas on which all the letters are based.
C. The remaining letters outline the author's sociopolitical thought.
D. The tenth letter presents guidelines for targeted action in light of the global process.

Here is a brief summary of the principal themes in this collection of letters:

First Letter: The situation in which we now happen to live; the disintegration of institutions and the crisis of solidarity; the new types of sensibility and behavior now beginning to take shape in today's world; criteria for action.

Second Letter: Change and related factors in today's world and the positions commonly taken in facing this change.

Third Letter: The characteristics of change and crisis in relation to the immediate environment in which each person lives.

Fourth Letter: The foundation of the views presented in these letters regarding the most general questions of human life, needs, and basic projects; the natural and social worlds; the concentration of power; violence and the State.

Fifth Letter: Human freedom, intention, and actions; the ethical meaning of social action and activism and their most common defects.

Sixth Letter: The ideological foundation of Humanism.

Seventh Letter: Social revolution.

Eighth Letter: The armed forces.

Ninth Letter: Human rights.

Tenth Letter: The phenomenon of generalized destructuring; applying global understanding to minimum concrete action.

The author has further explored the themes of the fourth letter, so important to the ideological underpinnings of all the letters, in the work *Contributions to Thought*, particularly in the essay entitled "Historiological Discussions," and also in the talk "The Crisis of Civilization and Humanism," which he presented at the Moscow Academy of Administration on June 18, 1992.

The sixth letter develops the central ideas of contemporary humanism. In its dense concentration of concepts this writing recalls political and cultural writings such as the manifestos of the last half of the nineteenth and early twentieth centuries, the Communist Manifesto or the Surrealist Manifesto, for example.

The use of the term *document* in place of *manifesto* in the sixth letter reflects a careful choice of words intended to distance the present Humanist Document from the naturalism embodied in the Dewey-inspired Humanist Manifesto I of 1933, and from the social-liberalism of Humanist Manifesto II of 1974, endorsed by Sakharov and strongly imbued with the thought of Lamont. Although there are similarities between Humanist Manifesto II and the present Document regarding the need for an economic and environmental plan that will not jeopardize personal freedom, there are radical differences between them in both their political visions and their conceptions of the human being.

The sixth letter deserves further comment in view of its extreme brevity in relation to the number of themes with which it deals. In it the author recognizes the contributions of various cultures to the development of humanism, as can clearly be observed in Jewish, Arabic, and Eastern thought. In this sense, this Document cannot be strictly placed in the Ciceronian tradition, as has usually been the case with Western humanisms. In recognizing historical humanism, the author has revived themes previously expressed as early as the twelfth century. I am referring to the goliard poets such as Hugh of Orléans and Peter of Blois, whose writings form part of the celebrated *Codex Buranus* (or Code of Beuren, also known as the "Songs of Bueren" and in Latin as *Carmina Burana*). Though Silo does not quote them directly, the following passages from his sixth letter recall their words:

> This is the great universal truth: Money is everything. Money is government, money is law, money is power. Money is basically sustenance, but more than this it is art, it is philosophy, it is religion. Nothing is done without money, nothing is possible without money. There are no personal relationships without money, there is no intimacy without money. Even peaceful solitude depends on money.

We hear a clear echo of the phrase from the *Codex Buranus*, "...the Abbot in his cell is kept a prisoner by money..." in Silo's phrase, "Even peaceful solitude depends on money." Or an echo of the line "Money receives honor, and none in want are loved..." in Silo's "There are no personal relationships without money, there is no intimacy without money."

The goliard poet's generalization, "Money, it is true, makes the fool seem eloquent," appears in this sixth letter as, "...but more than this it is art, philosophy, and religion." And of money the poem says, "Money is adored, for it makes such miracles...it makes the deaf hear and it makes the lame walk." In this poem from the *Codex Buranus* to which Silo alludes there is an implicit background that will later inspire the humanists of the sixteenth century, in particular Erasmus and Rabelais.

While the sixth letter presents the ideological foundation of contemporary humanism, I feel that an excellent way to provide an overview of this theme is to include here an excerpt from "A Contemporary Vision of Humanism," a talk given by the author at the Universidad Autónoma de Madrid on April 16, 1993.

> There are two meanings normally ascribed to the word *humanism*. Frequently people use the word humanism to mean any tendency of thought that affirms the dignity and

worth of the human being. Under this broad definition, humanism can be interpreted in the most diverse and contrasting ways.

In its more restricted meaning, placed within a precise historical perspective, the concept of humanism is used to indicate the process of transformation that began between the end of the fourteenth and the beginning of the fifteenth centuries, and that, known as the Renaissance, was to dominate the intellectual life of Europe over the next century. One need only mention the names of Erasmus, Giordano Bruno, Galileo, Nicholas of Cusa, Thomas More, Juan Luis Vives, and Charles Bouillé to illustrate the range and diversity of historical humanism. Its influence continued throughout the seventeenth and greater part of the eighteenth centuries, giving rise to the revolutions that ushered in the Modern Age.

The humanist current then seems to have slowly died out, until it was again rekindled toward the middle of this century in debate among thinkers concerned with social and political questions.

The broad outlines of historical humanism are in rough terms as follows:

1. Historical humanism forms a reaction against the Medieval way of life and values. Thus begins a strong recognition of other cultures, particularly the Greco-Roman, in art, science, and philosophy.
2. It introduces a new image of the human being that exalts the human personality and the transforming capacity of human action.
3. It embodies a new attitude toward nature, which now becomes accepted as the environment of humanity and no longer simply as a lower world filled with temptations and punishments.
4. There is a birth of interest in experimentation and research on the surrounding world, and a growing practice of seeking natural explanations without the need to invoke the supernatural.

These four aspects of historical humanism converge in the same objective: to galvanize a new confidence in human beings and in their creativity, viewing the world as the kingdom of humanity, a kingdom that can be mastered through a knowledge of the sciences.

From this new perspective one finds expressions of the need to construct a new vision of the universe and of history. At the same time, the new conceptions of this humanist movement spur a reevaluation of the religious question in its dogma and liturgy as well as in its organizational structures, which so permeate the social structures of the Middle Ages. Humanism, as a correlate of the changing economic and social forces of the period, represents a revolutionary new force that is increasingly conscious and oriented toward questioning the established order.

In response, the Reformation in the German and Anglo-Saxon worlds and the Counter Reformation in the Latin world attempt to halt the spread of the new ideas, reimposing the traditional Christian vision in an authoritarian way. The crisis spreads from the Church to the State structures. Finally the revolutions at the close of the eighteenth and in the nineteenth centuries bring to a close the age of empire and monarchy by divine right.

Then, following the French Revolution and the wars of independence in the Americas, Humanism all but disappears, although it persists as a social background of ideals and aspirations that fosters economic, political, and scientific transformations. Humanism retreats in the face of concepts and practices instituted with the close of the Colonial period, the Second World War, and the bipolar Cold War alignment of the world.

It is in this situation that we observe a reopening of debate regarding the meaning of the human being and nature, the justification of political and economic structures, the orientation of science and technology, and the general direction in which historical events are heading.

The first signs of this renewed debate appear among

Existential philosophers: Heidegger's rejection of Humanism as simply one more metaphysics in his "Letter on Humanism," Sartre's defense of it in his *Existentialism and Humanism,* and Luijpen in refining a theoretical framework in his *Phenomenology and Humanism.* Other noteworthy efforts were made by Althusser in the anti-Humanist position he presented in *For Marx* and by Maritain in appropriating humanism from its position as an antithesis of Christianity in *True Humanism.*

Having traveled this long road to the most recent discussions in the field of ideas, it is clear that Humanism today must define its position not only as a theoretical conception but also in terms of social action and practice. The state of the humanist question must today be articulated with reference to the conditions in which the human being actually lives. Such conditions are not abstract, and it is therefore not legitimate to derive Humanism from a theory of nature, a theory of history, or a faith in God.

The human condition is such that the immediacy of the encounter with pain and the need to overcome that pain are inescapable. This condition, common to so many other species, in the human being includes the additional need to foresee how in the future pain will be overcome and pleasure obtained. This human foresight is based on past experience and on the intention to improve present conditions. The work of human beings, accumulated in social productions, is transformed as it passes from one generation to others in the ongoing struggle to surmount the natural and social conditions in which they live.

Humanism therefore defines human beings as historical beings, whose mode of social action is capable of transforming both the world and their own nature. This point is of capital importance, for if it is accepted one cannot then affirm any natural law, natural property, natural institutions, or finally a future human being the same as that of today, as though the human being were complete now and forever.

Today the age-old theme of the relationship between the

human being and nature takes on renewed importance. On reconsidering it, we discover this great paradox in which human beings appear as without fixed character or nature, while at the same time we become aware of one constant in them: their historicity. This is why, stretching the terms a bit, we can say that **the nature of human beings is their history**, their social history. Each new human being who is born, then, is not simply equipped genetically to respond to the environment as though identical to the first representative of their species, but is rather an historical being whose personal experience unfolds within an accumulating social landscape, within a human landscape.

Yet we find in this social world that the intention we all hold in common – to overcome pain – is being negated by the intention of other human beings. What I am saying is that there are some human beings who naturalize others by negating their intention, converting them into objects to be used.

The tragedy of being at the mercy of natural physical conditions impels human social effort and science toward new achievements to surmount these conditions. In the same way, the tragedy of being subject to social conditions of inequality and injustice impels the human being to rebel against this situation, in which we find not blind forces at work but rather the interplay of other human intentions.

Those human intentions that negate the intention of others, that discriminate against other human beings, are to be judged within a wholly different context from that used to judge a natural tragedy, where there is no intention. This is why, in every act of discrimination, a monstrous effort is always made to establish that the differences between human beings are given by nature, whether physical or societal, that the differences at any rate are established by the interplay of blind forces, devoid of human intentions. So it is that racial, ethnic, sexual, economic, or other distinctions are made, and then justified by "laws," whether genetic or market. In every case, however, those who discriminate against others have to rely on distortion, untruths, and bad faith.

To sum up the two basic ideas outlined above, which frame the state of the question for today's humanists, first is the idea of the human condition as subject to pain, with the corresponding impulse to overcome this pain, and second is the definition of the human being as a social and historical being.

The foundational Document of the Humanist Movement (see Sixth Letter p. 69) observes that we will pass from pre-history into a truly human history when the violent animal appropriation of some human beings by others ceases. In the meantime one cannot begin from any other central value than that of the human being, fully realized and free. This is synthesized in the following declaration: "Nothing above the human being, and no human being beneath any other." If God, the State, money, or any other entity is placed as the central value, this subordinates the human being, creating the condition for the subsequent control or sacrifice of others.

As humanists we have this point very clear. We may be atheists or believers, but we do not begin with our atheism or our faith as the starting point for our vision of the world or for our actions. We begin with human beings and their immediate needs. As humanists, we state the fundamental problem as follows: to know if one wants to live, and to decide in what conditions to do so. All forms of violence – physical, economic, racial, religious, sexual, ideological, and so on – that have been used to restrain human progress are repugnant to humanists. Every form of discrimination, whether subtle or overt, is something that humanists denounce.

The following issues, then, mark a clear dividing line between Humanism and anti-Humanism: Humanism puts labor before big capital, real democracy before formal democracy, decentralization before centralization, anti-discrimination before discrimination, freedom before oppression, and meaning in life before resignation, complicity, and the absurd.

Because Humanism is based on freedom of choice, it offers a valid ethic; because Humanism believes in intention and freedom, it distinguishes between error and bad faith. This is how humanists define positions. We see ourselves not as having come out of nowhere, but as tributaries of a long process and collective effort. We are committed to our own times, even as we envision a long struggle that extends toward the future.

We affirm diversity, in direct opposition to the regimentation imposed until now. This regimentation relies on explanations that it is diversity itself that places the elements of a system in dialectic, so that to exercise respect for every particularity will unleash centrifugal forces of disintegration. Humanists hold exactly the opposite point of view, emphasizing that it is precisely the subjugation of diversity in this time that is leading to conditions that will prove explosive for rigid structures. This is why humanists **place the accent on an emerging convergence of direction, a convergence of intention, and why we oppose the idea and the practice of eliminating supposedly dialectic conditions within a given human group.**

Here ends our excerpt from this talk by the author.

The tenth and final letter establishes the limits of destructuring, focusing among many possible areas on three forms of destructuring in which this phenomenon takes on special importance: the political, the religious, and the generational. The letter also warns of the resurgence of fascist, authoritarian, and violent neo-irrationalist trends. In illustrating the idea of combining a global understanding with initiating action in the minimum sphere of one's immediate environment, the author makes a phenomenal leap of scale in which as readers we find ourselves back beside our neighbor, our coworker, our friend.

The letter makes a clear proposal that activists reject the mirage of superstructural political power because such power has been fatally wounded in the ongoing process of destructuring. In the future it will no longer matter who is president, prime minister, senator, representative, or deputy. Political parties, unions, and labor organizations will continue to drift ever further from their human base. The State will undergo a thousand transformations, and global decision-making power will be concentrated in only the largest of corporations and international finance capital, until finally this whole system is overtaken by the collapse of the para-State.

What, then, would be the value of an activism that strives to occupy what are only empty shells within a democracy that is merely formal? Clearly, actions must be proposed within the immediate environment of each person, since it is only from there, based on concrete conflicts, that real representation can be built. But the existential problems of the social base are not expressed exclusively as economic and political difficulties, and therefore a party that espouses a humanist ideology and instrumentally holds elected offices may have institutional significance, but will still be unable to meet the real needs of the people.

It is from the social base and as a broad movement, decentralized and federative, that new power will be constructed. The questions that all activists need to ask themselves are not "Who will be the next president or prime minister, the next representative or deputy?" but rather "How can we organize our centers for direct communication, our network of neighborhood councils? How can we give participation to all the organizations of the social base, no matter how small, through which people express their work, sports, art, culture, and religiosity?"

This Movement should not be conceived of in formal political terms, but instead as a process of converging diversity. Nor should the growth of this Movement be conceived of within the old mold of a gradualism that will supposedly gain space and social strata a little at a time. It should instead be proposed in terms of a "demonstration effect," characteristic of a multi-connected planetary society that is quite capable of reproducing and adapting a successful model in social environments that are very distant and different from one another. In sum, this final letter outlines a minimum type of organization and a strategy of action that correspond to the situation of today's world.

While I have limited my comments in this introduction to the fourth, sixth, and tenth letters in the belief that they have called for further references, citations, and complementary commentaries, I urge the reader not to overlook the many other letters, which do such a remarkable job of coherently tying together the broad range of contemporary humanist concerns, including, for example, human rights which are considered in the ninth letter, the current role of the armed forces treated in the eighth letter, the current state of activism discussed in the fifth letter, as well as each individual's personal ethical choices and possible coherence amid today's turbulent events which is considered in the third letter.

J. Valinsky
January, 1994

First Letter to My Friends

February 21, 1991

Dear Friends,

For some time now I have been receiving correspondence from various countries requesting that I explain or elaborate on certain of the subjects addressed in my books. For the most part what they have sought are explanations about such concrete issues as violence, politics, the economy, the environment, as well as social and interpersonal relationships. As you can see, these concerns are many and varied, and it is clear that the answers will have to come from specialists in these fields, which of course I am not. Yet while trying as far as possible not to repeat what I have written elsewhere, hopefully I will be able to present a brief outline of the general situation in which we are now living, along with some of the principal trends looming on the horizon.

In other eras, a certain idea of "cultural malaise" has been used as the unifying thread in this type of description. Here, in contrast, I will focus on the rapid changes taking place in the economies of different countries, as well as in their customs, ideologies, and beliefs, in an attempt to trace the particular type of disorientation that today seems to be asphyxiating both individuals and entire peoples.

Before entering the subject at hand, I would like to remark on two points. The first has to do with the world that has disappeared – a subject that may seem to some to be treated with a certain nostalgia in this letter. I will say on this point that those of us who believe in human evolution are not in the least depressed by the changes we see. On the contrary, we would like to see events accelerate faster still as we try to adapt ourselves increasingly to these new times.

The second point concerns the style of this letter – a style some may interpret as completely lacking in nuance, presenting these themes as it does in such a "primitive" way – so unlike the formulations of those whom we criticize. Regarding the form of expression that these champions of the "New World Order" might prefer, I simply offer the following comment. When speaking of these people, passages from two very different literary works keep echoing in my mind, George Orwell's *1984* and Aldous Huxley's *Brave New World*. Each of these exceptional writers foresaw a future world in which, through means either violent or persuasive, the human being is finally overwhelmed and reduced to an automaton. But I believe that, influenced perhaps by an undercurrent of pessimism which I will not attempt to interpret here, both writers in their novels attributed rather too much intelligence to the "bad guys" and too much stupidity to the "good guys."

Today's "bad guys" are very greedy people who have many problems, but who are in any case wholly incompetent to orient historical processes, processes that clearly elude both their will and their capacity to plan. These people, who are not very studious, are served in turn by technicians who possess only fragmentary and woefully inadequate resources. So I will ask you not to take too seriously those few paragraphs in which I have amused myself by putting in their mouths words they have not actually spoken, although their intentions do indeed go in the direction indicated. I believe that these matters should be approached without the customary solemnity so characteristic of this dying age, that instead they should be treated with the irreverent good humor one finds in letters exchanged between true friends.

1. The Present Situation

From the beginning of history, humanity has evolved through working to achieve a better life. Yet today, across wide regions of the planet, and in spite of the enormous advances achieved by humankind, what we see are power, economic might, and technology being used to murder, impoverish, and oppress people – destroying, moreover, the future of the generations to come and the overall equilibrium of life on this planet. While a tiny percentage of humanity now possesses great wealth, for the majority even their basic needs remain unmet. While in certain areas there may be sufficient jobs and adequate wages, in many other areas the situation is disastrous. And everywhere the most humble sectors of society undergo horrors each day simply to avoid starvation.

Today, and solely by the fact of having been born into a social environment, every human being should have access to an adequate level of nutrition, health care, housing, education, clothing, and services. And when they reach an advanced age, all people need to have a secure future for the remaining years of their lives. People have every right to desire these things for themselves, and they have every right to want their children to have a better life. But today, for thousands of millions of people, even these basic aspirations remain unfulfilled.

2. The Alternative of a Better World

Numerous economic experiments have been tried, with mixed results, in attempts to moderate the aforementioned problems. Today's trend is to apply a system in which we are told that hypothetical "market laws" will automatically regulate social progress, avoiding in this way the economic disasters of the previous experiments in controlled economies. According to this scheme, wars, violence, oppression, inequality, poverty, and ignorance will all fade away without any untoward consequences. Countries will integrate into regional markets, until finally we arrive at a global society that is without barriers of any kind. In this way, we are assured, just as the standard of living for the poorer sectors of developed regions will rise, so too will the less advanced areas receive the benefits of this progress.

The majority of people will adapt to this new arrangement, which competent technicians and business people will set in motion. If, however, something should fail to work out, it will certainly not be because of any problem with these infallible "natural economic laws," but only because of the shortcomings of those particular specialists – who, as happens in business, will simply be replaced as often as necessary. At the same time, in this "free" society the public will choose democratically among different options, always provided, of course, that their choices lie within this same system.

3. Social Evolution

Given the present circumstances, it is perhaps worthwhile to briefly reflect on this alternative, which is currently touted as the way to achieve a better world. Indeed, a great many economic experiments have been tried, yielding rather inconsistent results. Yet notwithstanding this, we are nonetheless being told that this latest experiment holds the only solution to our fundamental problems. There are, however, certain aspects of this new proposal that some of us fail to grasp.

First, there is the question of economic laws. It could appear plausible that, as in nature, there are certain mechanisms that through their free interplay will automatically regulate social evolution. However, we find great difficulty in accepting the argument that any human process, and certainly the economic process, belongs to the same order as natural phenomena. On the contrary, we believe that human activities are non-natural, that they are instead intentional, social, and historical. These particularly human phenomena do not exist in nature in general or in other animal species. Then, since economic processes reflect human intentions and interests, in light of events we see nothing to support the belief that those with control over the well-being of humanity are concerned with overcoming the difficulties of others less privileged than themselves.

Second, the assertion that societies have progressed notwithstanding the vast economic differences that have always separated the few "haves" from the majority of "have-nots" seems quite unsatisfactory. History demonstrates that peoples have advanced when they have demanded their rights from the established powers, and that social progress has clearly not been the result of some automatic "trickle down" of the wealth accumulated by one sector of society.

Third, it seems rather excessive to hold up as models certain countries that by operating within this so-called free market economic system have achieved a high standard of living. These countries have, after all, undertaken wars of expansion against other countries. They have imposed colonial and neo-colonial systems. They have partitioned nations and entire regions. They have exacted tribute through methods based on violence and discrimination. Finally, they have taken advantage of cheap labor in weaker economies, while at the same time imposing unfavorable trade terms on them. Some will argue that these procedures are no more than what are known as "good business deals." However, they cannot affirm this and then still claim that the economic development of these "advanced" countries has taken place independent of a special and unequal type of relationship with other countries.

Fourth, we frequently hear of the scientific and technical advances and the initiative that "free market" economies foster. But it is clear that scientific and technical progress began from the moment human beings invented clubs, levers, fire, and so forth, and that this progress has continued in a process of historical accumulation that has paid little heed to any particular economic form or set of market laws.

If, on the other hand, what they are trying to say is that the wealthy economies attract the largest part of the supply of talented people, that they have the resources to pay for equipment and research, and finally that they can provide more motivation in the form of greater compensation, then it should also be noted that this same phenomenon has occurred since ancient times, and is neither limited to nor the result of any one type of economy. Rather, it is simply that in this particular time and place – independent of the origin of such economic strength – an abundance of resources has accumulated.

Fifth, there remains the expedient of explaining the progress of "advanced" communities as the result of certain intangible natural "gifts" – special talents, civic virtues, hard work, organization, and the like. This is, however, no longer a rational argument, but instead a kind of devotional affirmation that, with some sleight of hand, obscures the social and historical realities that explain how those peoples were formed.

There are many of us, of course, who lack sufficient understanding to see how, given its historical background, the present market scheme will be able to survive even in the short run. But that forms part of another discussion – one that includes the question of whether this "free market economy" really exists at all, or whether in reality we are perhaps dealing with various forms of protectionism and indirect or disguised control, through which those in charge promptly loosen the reins in those areas where they feel in control and tighten them in areas where they do not. If this is the case, then every new promise of progress will remain in practice limited solely to the explosive development and spread of science and technology, which is independent of any supposed automatism in economic laws.

4. Future Experiments

Today, as throughout history, whenever necessary the prevailing scheme will simply be replaced by another that supposedly "corrects" the defects of the previous model. But all the while wealth will continue to concentrate step by step in the hands of an increasingly powerful minority.

At the same time, it is clear that neither evolution nor the legitimate aspirations of the people will come to a stop. So it is that soon we will see the last of any naive assurances that the end of ideologies, confrontations, wars, economic crises, and social unrest is at hand. And since no point on Earth is unconnected to the rest, both local solutions as well as local conflicts now rapidly become global. One other thing is certain: That which has prevailed until now can no longer be maintained – neither the present schemes of domination nor the formulas for struggle against them.

5. Change and Relationships Among People

The regionalization of markets, like the demands for local and ethnic autonomy, underscore the disintegration of the nation state. The population explosion in poorer regions is stretching to the breaking point all attempts to control migration. The large extended rural family is fragmenting, displacing younger members toward the overcrowded cities. The urban industrial and post-industrial family has shrunk to the minimum, while at the same time the macro-cities must absorb an enormous influx of people who were formed in disparate cultural landscapes. Economic crises and the conversion of productive models are giving rise to renewed outbreaks of discrimination.

In the midst of all this, technological acceleration and mass production result in products that are obsolete almost before they reach consumers. This continuous turnover of objects has a correspondence in the instability and dislocation so visible in contemporary human relationships. By now, traditional "solidarity," heir to what was once known as "fraternity," has lost all meaning. Our companions at work, school, in sports...even old friends have all taken on the character of competitors. Within couples, both partners struggle for control, calculating from the beginning of the relationship whether they have more to gain by staying together or separating.

Never before has the world been so closely interconnected, yet each day individuals experience a more anguishing lack of communication. Never before have urban centers been more populous, yet people speak of their "loneliness." Never before have people needed human warmth so much as now, but any approach to another in a spirit of kindness and help elicits only suspicion. This is the predicament to which our hapless people been abandoned, each isolated individual being led to believe in the greatest unhappiness that he or she has something important to lose – an ethereal "something" that is coveted by all the rest of humanity! Under such circumstances, the following story may be related as if it reflected the most authentic reality.

6. A Tale for Aspiring Executives

"The society now being set in motion will at last bring us prosperity. But apart from the enormous objective benefits, there will also be a subjective liberation of humanity. Old-fashioned 'solidarity,' a notion proper to poverty, will no longer be necessary, for by now practically everyone agrees that you can solve almost any problem with money, or its equivalent. We will therefore dedicate all our efforts, thoughts, and dreams toward this end. With money, you can buy fine food, a nice home, and afford travel, entertainment, high tech playthings, and people to carry out your wishes. At last there will be efficient love, efficient art, and efficient psychologists to correct any personal problems that remain. And soon, even these problems will be resolved, thanks to new developments in neurochemistry and genetic engineering.

"In this society of abundance we will see suicide, alcoholism, drug addiction, crime, and all those other insecurities of the urban dweller simply fade away – as is sure to happen any day now, we are assured, in the economically developed countries. Discrimination will disappear as well, and communication among all people will increase. No longer will anyone have to bear the sting of needless rumination on the meaning of life, loneliness, sickness, old age, or death, because, with the appropriate courses and a little therapeutic help, it will be possible to block these sorts of reflections that until now have been such a hindrance to society's output and efficiency. Everyone will trust everyone else, because competition at work, school, and in personal dealings will result in mature relationships.

"The last of the ideologies will finally disappear and no longer be used to brainwash people. Of course, no one will interfere with protest or nonconformity about minor things, provided that people express themselves through the appropriate channels. As long as they do not confuse liberty with license, citizens may gather (in small numbers, for reasons of hygiene), and may even express themselves outdoors (provided that they do not disturb others with noise pollution or publicity materials that could deface the municipality, or whatever it will be called in the future).

"The most extraordinary thing of all, however, will come to pass when police surveillance is no longer necessary, because every citizen will have resolved to protect others from the lies that could be inculcated by some dangerous ideological terrorist. On encountering suspicious activity, these guardians of the public welfare will rush to the news media, where they will find a warm welcome, and a warning will quickly be issued to the public. But the activities of these responsible citizens will not end there, for they will write brilliant studies, which will be published immediately. They will organize forums in which experts and pundits who shape public opinion will elucidate these things for the unwary, who would otherwise be at the mercy of the dark forces of state economic control, authoritarianism, anti-democracy, and religious fanaticism.

"It will, moreover, hardly be necessary to pursue these troublemakers. With such an efficient information system in place, no one will dare go near these dangerous elements for fear of being contaminated.

"The more serious cases will be efficiently 'deprogrammed,' and will publicly express their gratitude at being reintegrated into society and for the benefits they have received on recognizing the gifts of freedom.

"As a result of all this, those diligent guardians who have warned the public – if they were not sent specifically to carry out this vital mission – will be able to emerge from their anonymity and sign autographs as they attain the social recognition that befits their high moral character and, as is only logical, receive a well-deserved reward.

"The Company will be one big happy family, assisting with all phases of education, relationships, and recreation. Thanks to robots and automation, physical labor will no longer be required, and working for the Company from one's own home will provide genuine personal fulfillment.

"As a consequence, society will no longer have any need for organizations aside from the Company. Human beings, who have struggled for so long to achieve well-being, will at last reach the heavens – leaping from planet to planet they will discover true happiness. And that is where we will find our young citizen: well-established, competitive, charming, acquisitive, triumphant, and pragmatic – above all pragmatic – an executive in the Company!"

7. Human Change

The world is changing at a dizzying pace, and people can no longer hold on to much of what they believed unquestioningly until now. The acceleration of events is generating instability and disorientation in every society, rich and poor alike. In this situation of change, both traditional leaders and their "formers of public opinion," as well as old political and social activists, no longer serve as points of reference for people.

Yet a new sensibility is being born that corresponds to these changing times. It is a sensibility that grasps the world as a whole – an awareness that the problems people experience in one place involve other people, even if they are far away. Increasing communication, trade, and the rapid movement of entire human groups from one place on the planet to another all attest to this growing process of globalization.

As the global character of more and more problems comes to

be understood, new criteria for action arise. There is an awareness that the work of those who desire a better world will be effective only if they make their efforts grow outward from the environment where they already have some influence. In sharp contrast to other times, so full of empty phrases meant only to garner external recognition, today people are beginning to find value in humble and deeply felt work, work done not to enhance one's self-image, but rather to change oneself and bring about change in one's immediate environment of family, work, and friendship.

Those who truly care for people do not disdain this work done without fanfare, this work that proves so incomprehensible to those opportunists who were formed in an earlier landscape of leaders and masses – a landscape in which they learned well how to use others to catapult themselves to society's heights.

When a person comes to the realization that schizophrenic individualism is a dead end, when they openly communicate what they are thinking and what they are doing to everyone they know without the ridiculous fear of not being understood, when they approach others not as some anonymous mass but with a real interest in each person, when they encourage teamwork in both the interchange of ideas and the realization of common projects, when they clearly demonstrate the need to spread this task of rebuilding the social fabric that others have destroyed, when they feel that even the most "unimportant" person is of greater human quality than some heartless individual whom circumstance has elevated to what is, for now, the pinnacle of success – when all this happens it is because within this person destiny has once again begun to speak, the destiny that has moved entire peoples along their best evolutionary path, the destiny that has been so many times distorted and so many times forgotten, but is always rencountered in the twists and turns of history.

Today we can glimpse not only a new sensibility and a new mode of action but also a new moral attitude and a new tactical approach to facing life. If I were pressed to be more specific I would simply reply, though it has been said time and again over the last three millennia, that today people are experiencing anew the need for and the true morality of treating others as they want to be treated. I could add to this, almost as general laws of conduct, that today people are aspiring to:

1. A certain *proportion*, in which one tries to give order to the most important things in one's life, dealing with them as a whole and not allowing some aspects to move ahead while others fall too far behind.
2. A certain *growing adaptation*, in which one acts in favor of evolution rather than momentary concerns, turning away from the various forms of human involution.
3. A certain *well-timed action*, in which one retreats when facing a great force (not every little obstacle) and advances when that force weakens.
4. A certain *coherence*, in which one accumulates those actions that bring one a feeling of unity, of being in agreement with oneself, and rejects those actions that generate contradiction, that are registered within oneself as disagreements among what one thinks, feels, and does.

I do not feel it is necessary to elaborate on why I say that people are feeling anew "the need for and the true morality of treating others as they want to be treated," although some may object that this is not in fact how people act today. Nor do I believe it necessary to give lengthy explanations about what I understand by "evolution" or by "growing adaptation" as opposed to adaptation based on permanence. Concerning the parameters for knowing when to retreat or advance before a great or weakening force, people will certainly need to be able to recognize precise indicators beyond those mentioned here. Finally, it is obviously not easy to implement the proposal of accumulating unifying actions or, from the opposite point of view, rejecting contradictions, when dealing with the contradictory situations that touch our lives.

All of these considerations may be true, but if you review this letter you will see that these things have been discussed within the context of a new type of conduct to which people are today beginning to aspire – a type of behavior quite different from that to which people aspired in other times.

In this letter I have tried to note those special characteristics we see beginning to take shape that embody this new sensibility, this new type of personal conduct, and this new form of interpersonal action – all of which, it seems to me, go beyond a simple critique of today's situation. And while we know that criticism is always necessary, how much more necessary is it to do things in a new way – a way that is different from that which we criticize!

With this letter I send my

Warmest regards,
Silo

Second Letter to My Friends

December 5, 1991

Dear Friends,

In the previous letter I focused on the situation in which we now live and on certain tendencies visible in contemporary events. I also used the opportunity to discuss various proposals that defenders of market economics proclaim as if these were the inescapable preconditions for all social progress. I made note of the continuing decline in solidarity and the crisis of references now taking place. Finally, I outlined some positive characteristics that are beginning to appear in what I called a new sensibility, a new moral attitude, and a new tactical approach to facing life.

Some of my correspondents have expressed their disapproval of the tone of that letter, feeling it touched on subjects that are too grave to allow such irony. But let's not be so melodramatic – the system of proofs presented to justify the ideology of neoliberalism, social market economics, and the New World Order is so riddled with inconsistencies that this is hardly something to get worked up about.

I would like to point out that while the foundations of that ideology have long been dead, soon that entire edifice of ideas will be overtaken by a crisis so evident that even those who confuse meaning with expression, content with form, and process with circumstance will finally perceive it. Just as the ideologies of fascism and real socialism died long before these systems collapsed in practice, so too will the right-thinking people of today be caught by surprise as they recognize the collapse of the present system only after the fact.

Doesn't this all seem a bit ridiculous? It's like sitting through the same bad movie time after time. As we watch it over and over we begin to scrutinize tiny details – imperfections in the walls of the movie sets, the camera angles used, and whether the actors have shaved carefully – while the lady sitting beside us is overcome with emotion at what she is seeing for the first time, and what, for her, is reality itself.

On my own behalf, then, I might point out that I have not mocked the enormous tragedy that stems from the imposition of the present system, but instead the monstrous pretensions and grotesque end of this system – an ending that we have already witnessed before on too many previous occasions.

I have also received correspondence requesting more precise definitions of the attitudes recommended for facing the present process of social change. Before making any recommendations of this kind, however, I believe it would first be useful to try to understand the principal positions now held by various groups as well as by isolated individuals. Here I will limit myself to presenting the most popular positions, giving my views in those cases that seem to be of greatest interest.

1. Some Positions Regarding the Present Process of Change

Throughout the long ascent of humanity progress has occurred in a slow process of accumulation up to the present time, when the pace of economic and technological change has begun to outstrip the speed of change in social structures and human behavior. Many factors in society are becoming more "out of phase" all the time, which is generating growing crises in today's world.

This problem can be approached from various points of view. Some believe that the current disarticulation will automatically regulate itself, and they therefore recommend that we not attempt to direct this process, which would in any case be impossible to orient. This approach embodies an optimistic-mechanistic thesis. Still others believe we are heading toward an inevitable explosion – they hold a pessimistic-mechanistic thesis. Various moral currents are also making their appearance, attempting to stop change and, as far as possible, return to some original past where they assume that comfort is still to be found. They represent an anti-historical position. Meanwhile, all around us we hear a rising chorus of voices from contemporary cynics, stoics, and epicureans. The first deny that there is importance or meaning in any action at all. The second face events unflinchingly, even when everything goes badly. Those who adopt the third position seek personal benefit in every situation, thinking only of their own hypothetical well-being, which extends, at most, to their own children.

As in the final stages of past civilizations, many people today are opting for positions that pursue individual salvation, assuming that no task they might undertake with others could have any meaning or possibility of success – at most others have a useful role to play only insofar as they profit one within a speculation that is strictly personal. That is why aspiring business, cultural, and political leaders perfect and polish their public images, striving to seem credible so that people will believe they think of and act on behalf of others. This is, of course, a rather fruitless task, because by now everyone knows the tricks and no one believes in anyone else.

The old values – religious, patriotic, cultural, political, union, and so on – have all been subordinated to money in a landscape in which solidarity and, therefore, any collective opposition to the contemporary scheme of things has been eroded, even as the social fabric continues to unravel. Afterwards, another stage will follow in which this inordinate individualism will be outgrown – but that is a theme for later on.

With our landscape of formation weighing us down and our beliefs in crisis, we are not yet in any condition to admit that this new historical moment is approaching. Today, whether we wield some small measure of power or depend absolutely on the power of others, we all find ourselves touched by this individualism – a situation in which those who are better placed in the system have a clear advantage.

2. Individualism, Social Fragmentation, and the Concentration of Power in a Few

Individualism necessarily leads, however, to the struggle for the supremacy of the strongest and the pursuit of "success" at any price. This position began among a few who, relying on the acquiescence of the majority, respected certain rules of the game among themselves. In any event, this stage will soon exhaust itself and it will become "all against all," because sooner or later the balance of power will tilt in favor of the strongest, and then the rest, either together or in alliances of various factions, will end up dismantling this fragile system.

In the meantime, however, as economies and technologies continue to develop, the powerful minorities continue to change along with them, perfecting their methods to such a degree that in some wealthy areas the majorities now effectively transfer their discontent to secondary aspects of the predicament in which they live. It appears that people generally no longer question the system as a whole but only certain urgent aspects when these strike close to home. Because of this, there are some who suggest that despite the overall rise in the world's wealth and standard of living, the great masses of humanity who are left behind will simply be content to await a better life in some distant future.

All of this demonstrates an important shift in social behavior. And if this has occurred, activism for social change will continue to weaken as traditional political and social forces are left devoid of proposals. With the emptiness of individual isolation only partially filled by those structures that produce goods and leisure activities, the fragmentation of personal and collective life will continue to increase.

In this paradoxical world, all centralization and bureaucracy will be swept aside, breaking with the former structures of management and decision making. Yet at the same time, this deregulation, decentralizing, and liberalizing of markets and procedures will leave the field wide open for the concentration of wealth and power to flourish on a scale unknown in any previous era, as international finance capital continues to flow into the hands of an ever more powerful banking system.

The political class will experience a similar paradox in that they will have to champion these new values, which in eroding the power of the State will simultaneously undermine their own leadership role. It is little wonder then that for some time they have been replacing words such as "government" with other words such as "administration," trying to lead "the public" (no longer "the people") to understand that a country is now a business.

In any event, and until the consolidation of a global imperial power, conflicts between regions could well occur as previously they occurred among countries. Whether such confrontations will be limited to the economic sphere or spill over into the arena of limited warfare, whether massive and incoherent unrest will as a consequence erupt, whether governments will fall pulling down countries and whole regions, will not in the least deter the process of concentration toward which this historical moment is heading. Local grievances, inter-ethnic fighting, migrations, refugees, sustained crises – none of these will alter the general picture of the increasing concentration of power.

And when the recession and unemployment become chronic among the populations of the wealthy countries, the stage of liquidating any remaining liberalism will have finished, ushering in the politics of control, coercion, and emergency in the finest imperial style – and who then will be able to speak of a free market economy, and what importance will it have to maintain positions based on an uncompromising individualism?

In this letter I will also respond to other concerns that my correspondents have raised concerning how to characterize the current crisis and its associated tendencies.

3. Characteristics of the Crisis

Let us turn now to the crisis of the nation state, the crisis of regionalization and globalization, and the crisis facing society, the group, and the individual.

In the context of the process of globalization, the flow of information is accelerating as the movement of both people and goods continues to increase. Technology and growing economic power are becoming concentrated in businesses that are ever more powerful. And this phenomenon of accelerating interchange is now encountering the limitations and slowed pace that are produced by traditional structures such as the nation state.

The result is that within each region national borders are becoming blurred. This means that countries are having to make their legislation more homogeneous, not only in matters of trade regulations, duties and tariffs, and personal documentation, but also in adapting their systems of production. Changes in labor and social security laws cannot be far behind. Ongoing accords among these countries will show that a common legislature, judicial system, and executive will provide improved effectiveness and quicker response time in managing the region. Primitive national currencies will give way to some type of regional medium of exchange that will avoid the losses and delays of previous exchange operations.

The crisis of the nation state is a readily observable fact, not only in those countries that are joining to form regional markets but also in those whose battered economies have fallen significantly behind. Everywhere voices are being raised against entrenched bureaucracies, demanding the reform of established schemes. Old resentments as well as local, ethnic, and religious rivalries are resurfacing in regions where countries have recently been formed as a result of partitions, annexations, or artificial federations. And the traditional State is having to face this centrifugal tendency at just the time that growing economic difficulties are calling into question its effectiveness and legitimacy.

Phenomena of this type are growing in the areas of Eastern

Europe, the Balkans, and the former Soviet Union. These problems will also deepen in the Middle East, the eastern Mediterranean, and Asia Minor. In a number of countries of Africa that have been artificially delimited we are beginning to see such symptoms as well. Accompanying these breakdowns are large-scale migrations of refugees toward borders, which can threaten the equilibrium of an entire region. With any significant imbalance in China, this phenomenon could spill directly into more than one other area, especially in light of the present instability in the former Soviet Union and the countries of continental Asia.

In the meantime, the regional centers of economic and technological power have become configured: the Far East, led by Japan; Europe; and the United States. While the rise and influence of these regions exhibit an apparent polycentrism, events demonstrate that the United States with its military might in addition to its technological, economic, and political power is now in a position to control the world's key lines and areas of supply.

In the process of increasing globalization, this lone remaining superpower is emerging as the governing force in present events, whether the other regional powers like it or not. This is the ultimate meaning of the New World Order.

It seems that we have yet to reach a time of peace, although the threat of world war has receded for now. Local, ethnic, and religious upheavals, social unrest, mass migrations, and limited wars still appear to threaten the supposed present stability. As the less wealthy areas fall still further behind the growth of the technologically and economically accelerated areas, they become more "out of phase," which only compounds their problems. Latin America is a case in point, for even as the economies of various countries experience important growth in coming years, their dependence on the centers of power will be increasingly evident.

As the regional and world power of multinational companies continues to grow, as international finance capital continues to concentrate, political systems lose autonomy and their legislation must adapt to the dictates of these new powers.

Today we see the functions of increasing numbers of institutions being directly or indirectly supplanted by various departments or foundations of the Company, which in some areas has developed the means to oversee everything from cradle to grave for both employees and their children: birth, education, career placement, news and information, marriage, recreation, social security, retirement, death, and burial.

There are already places where citizens can avoid old-fashioned bureaucratic paperwork and get by with only a credit card and, increasingly, with just electronic money. And when people use electronic money, a record is made of not only their expenditures and deposits, but also of a wealth of other pertinent information on their background, habits, movements, present status, and so forth, all duly computerized. Of course, while this does free some people from a few minor delays and concerns, these personal conveniences also serve a disguised system of control. Along with the growth in technology and the accelerating rhythm of life, political participation diminishes and decision-making power becomes ever more remote and intermediated.

The family is shrinking and flying apart into the minimum unit of increasingly mobile and changeable couples. As interpersonal communication becomes blocked, friendship disappears, and competition poisons all human relationships to the point that no one trusts anyone else. The sensation of insecurity that people are feeling is no longer rooted in the objective fact of rising crime and violence, but stems above all from their state of mind. It must be added that social, group, and interpersonal solidarity are rapidly disappearing, that drug addiction and alcoholism are continuing to spread devastation, and that suicide and mental illness are spiraling dangerously upward. Of course, everywhere there is still a healthy and reasonable majority, but the symptoms of such advanced disarticulation no longer allow us to speak of a healthy society.

The landscape of formation in which the new generations have grown up contains all the elements of crisis previously cited, and these elements form part of their lives just as much as their technical and career training, as much as elements like soap operas, the advice of celebrity experts in the mass media, affirmations about what a perfect world we live in and, for more privileged youth, the diversions of motorcycles, travel, clothes, sports, music, and electronic gadgets. The problem of this landscape of formation in the new generations threatens to widen the already enormous gap between sectors of different ages, bringing to the fore a virulent generational dialectic of both great depth and vast geographical extension.

It is clear that the myth of money has long since been incorporated at the pinnacle of the scale of values, with everything else increasingly subordinated to it. A large segment of society does not want to hear about anything that could remind them of old age or death, shunning any theme related to the meaning and direction of life. And we must recognize that this is not altogether unreasonable, since reflection on these subjects in no way coincides with the scale of values established in the present system.

The symptoms of the crisis are by now too serious to disregard, yet some will maintain that this is simply the price we must pay in order to exist at the close of the twentieth century. Others affirm that we are entering the best of all possible worlds. The background for both of these affirmations comes from this particular historical moment, when the whole scheme of things has not yet entered crisis, although particular crises are proliferating rapidly. People's appreciation of events will change, however, as the symptoms of disintegration accelerate and they feel the growing need to establish new priorities and new projects in life.

4. Positive Factors of Change

One cannot question the entire development of science and technology simply because some advances have been or are being employed against life and the well-being of all. In any questioning of science and technology one must first reflect on the characteristics of the prevailing system, which all too often applies advances in knowledge toward spurious ends. Progress in medicine, communications, robotics, genetic engineering, and myriad other fields can of course be applied in a destructive direction. The same holds true of employing technology in the irrational exploitation of natural resources and the generation of industrial pollution, with attendant widespread contamination and deterioration of the physical environment. All such misuse of technology constitutes a grave indictment of the negative character that now commands both the economy and social systems.

Today it is clear that society has the capacity to solve the problems entailed in feeding all of humanity, and yet every day we see starvation, malnutrition, and inhuman suffering increase around us. In short, the established system is not disposed to face these problems and relinquish its fabulous profits in exchange for an overall improvement in the human condition and standard of living.

It must also be pointed out that the process carrying us toward regionalization and finally globalization is being manipulated by special interests to the detriment of humanity as a whole. It is clear, however, that even burdened with such distortions this process is opening the way toward a *universal human nation*. The accelerated change taking place in today's world is leading to a global crisis for the system and a consequent reordering of many factors. And all of this will be the necessary condition to reach a reasonable stability and harmonious development of the planet.

Accordingly, despite the tragedies that can be anticipated as the present global system deteriorates, the human species will prevail over all particular interests. This faith in the future is rooted in an understanding of the direction of history that began with our hominid ancestors. This species, which has worked and struggled over the course of millions of years to surmount pain and suffering, is not now going to yield to the absurd. This is why we need to understand processes that are more ample than simple immediate circumstance, and to support, even if we do not see immediate results, everything that goes in the direction of evolution.

When courageous human beings who are moved by a spirit of solidarity become disheartened it slows the march of history. But it is difficult to grasp this broader meaning if one does not also organize and orient one's personal life in a positive direction. What is at work here is not the interplay of mechanical factors or historical determinism – it is human intention, which tends to make its way through all difficulties.

I hope, my friends, to move on in the next letter to other more reassuring topics, leaving aside observations concerning such negative factors in order to outline proposals that correspond to our faith in a better future for all.

With this letter I send my

Warmest regards,
Silo

Third Letter to My Friends

December 17, 1991

Dear Friends,

I hope that this letter will help simplify and give order to my views on the present state of affairs. In it I also want to consider some important aspects of the relationships between individuals and between individuals and the social environments in which they live.

1. Change and Crisis

In this time of great change, individuals, institutions, and society all find themselves in crisis. And the pace of change – and the intensity of these individual, institutional, and social crises – will only continue to increase. This portends further upheaval, which broad sectors of society will perhaps be unable to assimilate.

2. Disorientation

Today's transformations are taking unexpected turns, resulting in widespread disorientation about the future and confusion about what to do in the present. In reality, it is not change itself that is so disturbing to us, because we can recognize many positive things in contemporary developments. What is troubling is not knowing in what direction these changes are heading, and therefore not knowing in what direction to orient our actions.

3. Crisis in the Life of Each Person

Everything around us – the economy, technology, society – is undergoing enormous transformations. But above all it is in our own lives that we experience these changes: in our workplaces, our families, our friendships, and not least in our ideas and what we believe about the world, other people, and ourselves. Amid the rush of events we find many things exciting, yet other things confuse or paralyze us. Our own behavior and that of others all too often seems incoherent, contradictory, and as lacking in any clear direction as the events around us.

4. The Need to Give Direction to One's Life

Since change is inevitable, it is of fundamental importance to guide it, and there is no other way than to begin with oneself. One must find in oneself a direction for this chaotic change, whose future course is unknown to us.

5. Direction in Life and Changing One's Situation

Individuals do not exist in isolation. Thus, if they truly give their lives direction, this will change their relationships with the people in their families, their workplaces, and everywhere they carry out their activities. Giving direction to one's life is not simply a psychological problem that can be resolved within the head of an isolated individual; on the contrary, it is resolved by changing – through coherent behavior – the situation in which one lives with others.

When we become excited by our successes or depressed by our failures, when we make plans for the future or resolve to change our lives, we often forget the fundamental point: The situation in which we live involves relationships with others. We can neither explain what happens to us nor make any choice in our lives without also including certain people and concrete social ambits. Those people who are of special importance to us and the social environments in which we live place each of us in a particular situation, and it is from this situation that each of us thinks, feels, and acts. To deny this or to disregard it creates enormous difficulties both for us and for others. One's freedom to choose and to act is delimited by these circumstances. Any change one desires to make cannot be proposed in the abstract but only with reference to the actual situation in which one lives.

6. Coherent Behavior

If my thoughts, my feelings, and my actions are in agreement, if they all go in the same direction, if my actions do not create contradiction with what I feel, then I can say that my life has coherence. But though I am true to myself, this does not necessarily mean I am being true to those in my immediate environment. I still need to achieve this same coherence in my relationships with others, treating them the way I would like to be treated.

Of course there can also be a destructive type of coherence, which can be seen in those who are racists or fanatics or in those who are violent or exploit others. It is clear, however, that **their relationships with others are incoherent**, because they treat others very differently from the way they desire to be treated themselves.

That unity of thought, feeling, and action, that unity between the treatment one asks from others and the treatment one gives to others – these are ideals that are not realized in everyday life. Here is the point: to adjust one's conduct in the direction of these personal and social proposals. These values, taken seriously, give life a direction that is independent of any difficulties one may face in realizing them. If we observe things well – not in static but in dynamic – we will understand this as a strategy that continues to gain ground as time passes. Here, one's intentions *do* matter (even though one's actions may at first not coincide with them), especially if these intentions are sustained, perfected, and extended. These images of what one wants to achieve are firm references that give direction in every situation.

What is being proposed here is not very complicated. We are not surprised, for example, when people dedicate their lives to pursuing great wealth, even when they lack any tangible reason to believe they will achieve it. This ideal spurs them on, despite the absence of relevant results. Why, then, is it so difficult to understand that these ideals of how to treat others and personal coherence can provide a clear direction for human conduct? And these ideals can give one direction despite the fact that these times are neither conducive to having the treatment one asks correspond to the treatment one gives nor to having one's thoughts, feelings, and actions be in agreement.

7. The Two Proposals: Coherence and Solidarity

To have one's thoughts, feelings, and actions go in the same direction and to treat others as one wants to be treated – these two proposals are so simple they can be viewed as mere naiveté by people accustomed to the usual complications. Yet underlying this seeming simplicity lies a new scale of values in which coherence comes first, a new morality in which one's actions are not a matter of indifference, and a new aspiration that entails a consistent effort to give direction to human events. Behind this apparent simplicity one is either staking one's future on a meaning in life that will be truly evolutionary, both personally and for society, or one is following a path that leads toward disintegration.

As mistrust, isolation, and individualism increase, they erode the fabric of society, and we can no longer rely on old values to provide the cohesion among people that is so essential. The traditional solidarity found among members of a given class, or within associations, institutions, and groups is rapidly being replaced by a savage competition, from which not even the closest bonds of marriage or family escape.

As this process mechanically proceeds to dismantle social structures, a new solidarity cannot arise out of the ideas and conduct of a world that has already disappeared – it can come only from the concrete need that people have to give direction to their lives. And this new direction will entail changing the environment in which they live. This change in their environment, if it is to be true and profound, cannot be imposed from without, cannot be set in motion by external laws or any form of fanaticism. It can only come from the power of shared opinion and minimum collective action with the people who make up the social environment around them.

8. Reaching All of Society Starting with One's Immediate Environment

We know that by changing our situation in positive ways we will be influencing our surroundings, and that others will share this point of view and form of action, giving rise to a growing system of human relationships.

So we must ask ourselves: Why should we go beyond the immediate environment where we begin? The answer is simple: To be coherent with the proposal of treating others in the same way we want them to treat us. Why wouldn't we pass on to others something that has proven to be of fundamental importance in our own lives?

If our influence begins to expand, it means that our relationships and therefore the constituents of our environment have also developed. This is a factor we need to bear in mind right from the first, because even though our actions may begin in one small area, their influence can project very far. And there is nothing strange in thinking that others will decide to accompany us in this direction. After all, the great movements throughout history have followed this same course – logically, they began small, and then developed because people felt these movements interpreted their needs and concerns.

If we are coherent with these proposals we will act in our immediate environments, but with our vision placed on the progress of society as a whole. For what meaning is there in speaking of a global crisis that must be faced with resolution if society is only going to end up as isolated individuals for whom others have no importance?

Out of common need, then, those working together to give a new direction to their lives and to events will create environments for direct communication where they can discuss these themes. Later on, as awareness spreads through many means of communication, this surface of contact will grow. A similar process will occur as people create organizations and institutions compatible with this proposal.

9. The Social Environment in Which One Lives

We have already seen that the impact of this swift and unpredictable change is experienced as crisis – the crisis with which individuals, institutions, and entire societies are now struggling. So, although it is indispensable to give direction to developments, how can one do this, subject as one is to the action of larger events? Clearly, one can direct only the most immediate and nearby aspects of one's life, and not the operation of institutions or society at large. Nor is it easy attempting to give direction to one's life, since no one lives in isolation; everyone lives in some situation, in some environment.

We may think of this environment as the universe, the Earth, our country, state, province, and so on. Each of us has, however, an immediate environment – the environment in which we carry out our daily activities. This is the environment of our family, our work, our friendships, and our other activities. We live in a situation of relationship with other people, and this is our particular world, which we cannot avoid as it acts on us and we on it in a direct way. Any influence we have is on this immediate environment, and both the influence we exercise on it and the influence it exerts on us are in turn affected by more general situations – by the current disorientation and crisis.

10. Coherence as a Direction in Life

If we want to give a new direction to events, we must begin with our own lives and include the immediate environment in which we carry out our activities. But the question remains: To what direction will we aspire? Without doubt to one that provides coherence and support in such a changeable and unpredictable environment.

To propose that one will think, feel, and act in the same direction is to propose coherence in life. Yet putting this into practice is not easy, because the situations in which we find ourselves are not entirely of our own choosing. We find ourselves doing the things we need to do, even though these things may not at all agree with what we think or what we feel. We find ourselves in situations over which we have no control. To act with coherence, then, is more an intention than a fact – it is a direction, which if kept before us guides our lives toward increasingly coherent conduct.

Clearly, it is only by exerting influence within one's own immediate environment that one will be able to change any aspect of the overall situation in which one lives. In so doing, one will be giving a new direction to one's relationships with others, and they will be included in this new conduct.

Some may object that their employment or other factors cause them to frequently change their residence or other aspects of their lives. But this in no way affects the proposal, for every person is always in some situation, is always part of some environment. If we are striving for coherence, the treatment we afford others must be of the same type as the treatment we demand for ourselves, no matter where we are.

There are, then, in these two proposals the basic elements for giving direction to our lives to the extent of our strength and possibilities. Coherence advances as a person is increasingly able to think, feel, and act in the same direction. And we extend this coherence to others – because only in this way are we ourselves being coherent. And in extending this to others we begin to treat other people the way we would like to be treated. Coherence and solidarity are directions, they represent conduct to which we aspire.

11. Proportion in One's Actions as a Step Toward Coherence

How can we advance in the direction of coherence? First, we need to maintain a certain proportion in the activities of our daily lives. We need to establish which among all the things we do are most important. For our lives to function well, we need to give the highest priority to what is of fundamental importance, less to secondary things, and so on. It could turn out that simply by taking care of two or three main priorities we will achieve a well-balanced situation.

We cannot allow our priorities to be turned upside down or to become so fragmented that our lives grow out of balance. To avoid having some activities proceed far ahead while others fall too far behind, we need to develop all of our activities as a connected whole and not as isolated actions. It is all too easy to become blinded by the importance of one activity and to allow this single priority to unbalance all of our other activities. And then, because our whole situation has been jeopardized, in the end we fail to accomplish what we had considered so important.

It is true that at times urgent matters arise that we need to deal with right away, but it should be clear that this in no way means we can go on indefinitely postponing the things necessary to maintain the overall situation in which we live. It is a significant step in the direction of coherence to establish our priorities, and then to carry out our activities in appropriate proportion.

12. Well-Timed Actions as a Step Toward Coherence

There is a daily routine we follow that is set by schedules and timetables, our personal needs, and the workings of the environment in which we live. Yet within this framework there is a dynamic interplay and richness of events that go unappreciated by superficial people. There are some who confuse their routines with their lives, but they are in no way the same, and quite often people must make choices among the routines or conditions imposed on them by their environment.

Certainly it is true that we live amid inconveniences and contradictions, but it is important not to confuse these things. *Inconveniences* are simply the annoyances and impediments that we all face. While they are not terribly serious, of course if they are numerous or repeated they can increase our irritation and fatigue. Without question we have the capacity to overcome them. They neither determine the direction of our lives nor stop us from carrying a project forward. They are simply obstacles along the way that range from the minor physical difficulty to larger problems that may nearly cause us to lose our way. While there are important differences in degree among inconveniences, they all lie within the range of things that do not stop us from going forward.

Something quite different happens with what are called *contradictions*. When we are unable to carry out our central project, when events propel us in a direction away from what we desire, when we find ourselves trapped in a vicious circle from which we cannot escape, when we do not have even minimal control over our lives, then we are ensnared by contradiction.

In the stream of life, contradiction is a sort of countercurrent that carries us backward in hopeless retreat. This is incoherence in its crudest form. In a situation of contradiction, one's thoughts, feelings, and actions oppose each other. And though in spite of everything it is always possible to give direction to one's life, one has to know *when* to act.

In the routine of daily life we often lose sight of whether or not our actions are timely, and this occurs because so many of the things we do are codified or set by convention. But when it comes to major difficulties and contradictions, we must not make decisions that expose us to catastrophe.

In general terms, what we need to do is to retreat when faced with a great force, and then advance with resolution when this force has weakened. There is, however, a great difference between the timid, who retreat or become paralyzed when faced with any difficulty, and those who take action to surmount the difficulties, knowing that it is precisely by advancing that they will be able to get through the problems.

At times it may happen that it is not possible to go forward immediately because a problem arises that is beyond our strength, and to tackle it head on without due care could lead to disaster. This problem we are facing that is now so large is also, however, dynamic, and the relationship of forces will change, either because our influence grows or because the problem's influence weakens. Once the previous balance of forces has shifted in our favor, that is the moment to advance with resolution, for indecision or delay at that point will only allow further and perhaps unfavorable changes in the balance of forces. Well-timed action is the best tool to produce a change in the direction of one's life.

13. Growing Adaptation as an Advance Toward Coherence

Let us further consider the theme of direction in life – of the coherence we want to achieve. To propose a direction toward coherence raises the question: To which situations should we adapt?

To adapt to things that lead away from coherence would, of course, be highly incoherent, and opportunists suffer from a serious shortsightedness on precisely this point. They believe that the best way to live is simply to accept everything, to adapt to everything. They think that to accept everything, as long as it comes from those with power, is to be well-adapted. But it is clear that their lives of dependence are very far removed from what could be understood as coherence.

It is useful to distinguish three kinds of adaptation: being *unadapted*, which stops us from extending our influence; *decreasing adaptation*, in which we do not go beyond accepting the established conditions in our environment; and *growing adaptation*, through which we build our influence in the direction of the proposals outlined here.

To close, let us synthesize the themes of this letter:

1. Driven by the technological revolution, the world is undergoing rapid change, which is colliding with established structures and the formative experience and habits of life of both individuals and societies.
2. As change makes more factors in society become "out of phase," this generates growing crises in every field, and there is no reason to suppose this will diminish; on the contrary it will tend to intensify.
3. The unexpectedness of today's events clouds our ability to foresee the direction that these events, the people around us, and ultimately our own lives will take.
4. Many of the things we used to think and to believe in no longer work. Nor do we see adequate solutions forthcoming from any society, any institution, or any individual – all of whom suffer the same ills.
5. If one decides to stand up to these problems, one must give direction to one's life, striving for coherence among one's thoughts, feelings, and actions. And because we do not live in isolation, we must extend this coherence to our relationships with others, treating them as we want to be treated. While it is not possible to fulfill these two proposals rigorously, nonetheless they constitute the direction in which we need to advance, which we will be able to accomplish above all if we make these proposals permanent references, reflecting on them deeply.
6. We live in immediate relationship with others, and it is in this environment that we must act to give a favorable direction to our lives. This is not a psychological question, a matter that can be resolved solely in the head of an isolated individual, it is related to the concrete situation in which each of us lives.
7. Being consistent with the proposals we are attempting to carry forward leads us to the conclusion that it would be useful to extend to society as a whole those elements that are positive

for ourselves and our immediate environment. Together with others who are moving in this direction, we will put into practice the most appropriate means to allow a new form of solidarity to find expression. Thus, even when we act very specifically in our own immediate environment we will not lose sight of the global situation that affects all human beings and that requires our help, just as we need the help of others.

8. The precipitous changes in today's world lead us to to seriously propose the need for a new direction in life.
9. Coherence does not begin and end in oneself, rather it is related to one's social environment, to other people. Solidarity is an aspect of personal coherence.
10. Proportion in one's activities consists of establishing one's priorities in life, of not letting them grow out of balance, and basing one's actions on these priorities.
11. Well-timed actions involve retreating when faced with a great force, and advancing with resolution when it weakens. When one is subject to contradiction, this idea is important in making a change of direction in one's life.
12. It is unwise to be unadapted to our environment, which leaves us without the capacity to change anything. It is equally unwise to follow a course of decreasing adaptation to an environment in which we limit ourselves to accepting the established conditions. Growing adaptation consists of increasing the influence we have in our environment as we advance in the direction of coherence.

With this letter I send my

Warmest regards,
Silo

Fourth Letter to My Friends

December 19, 1991

Dear Friends,

In previous letters I have given my views on society, human groups, and individuals in relation to this moment of change and loss of references in which we happen to live. I critiqued certain negative tendencies in the development of events and outlined the better-known positions held by those who claim to have answers to the urgent concerns of these times.

It should be clear that all of these considerations, whether well or badly formulated, correspond to my particular point of view, and this in turn finds its foundation in a certain set of ideas. No doubt due to an awareness of this on the part of some of my correspondents, I have received encouragement to make more explicit from what point of view, from "where," my critiques and proposals are developed.

After all, in the course of our daily lives ideas occur to us that may or may not be very original, but that in any case we don't claim to justify. And increasingly we find that we hold one idea today and the opposite one tomorrow, without going beyond the frivolity of an everyday appreciation of things. Each day, then, we believe less – not only in the opinions of others but even in our own – as we become accustomed to seeing opinions as something transient, changing from hour to hour as they fluctuate with the volatility of the stock market. And if among these varied opinions some do possess greater permanence, it is only because they are consecrated by the fashion of the day, which will always be replaced by the fashion of tomorrow.

I am not defending the value of unchanging opinions, I am simply pointing out the current lack of consistency among opinions generally. In truth it would be very interesting for changes in people's opinions to come about based on an internal logic and not simply as though bending before every erratic wind. But who today has any taste for internal logic, with so many flailing around as though drowning in these turbulent times. Even as I write this, I am keenly aware that what I say will not even be able to enter the heads of certain readers, because they will have failed to find one of the three possible codes they demand, which are:

1. That this letter provides them with entertainment; or
2. That this letter provides them with something they can use at once in their business; or
3. That this letter coincides with what is consecrated by fashion.

I am certain that these few paragraphs beginning with "Dear Friends" and extending to here will leave certain readers as thoroughly bewildered as if they were written in Sanskrit. Yet every day these same persons understand matters of great difficulty ranging from sophisticated banking operations to the exquisite niceties of computer network administration. Somehow, however, such people find it impossible to understand that in this letter I am speaking of opinions, of certain points of view, and of the ideas that serve as their foundation – and of the impossibility that they will understand even the simplest of these things if these matters do not correspond to the landscape they have assembled in the course of their educations and their compulsions. So this is how things stand!

Having addressed that question, I will now try to summarize in this letter the ideas that form the foundation of my views, critiques, and proposals. I will exercise care in presenting things not to go much beyond the level of advertising slogans because, as we are cautioned by many learned and expert journalists, organized ideas are "ideologies," and these, like doctrines, are today only instruments of brainwashing employed by those who oppose free trade and social economics in the marketplace of opinion, which these guardians so carefully regulate for our benefit.

Those people who conform to the demands of post-modernism today – who heed the requisites of *haute couture* with evening wear, flashy ties, shoulder pads, running shoes, and dapper jackets, who follow the dictums of deconstructionist architecture and destructured decor – demand of us that the elements of our discourse not fit together. And let us not forget that their critique of language repudiates as well all that is systematic, all that is structural, and everything related to processes!

Of course, it will come as no surprise that this position corresponds to the dominant ideology of the Company, in whose representatives there is a horror of history, just as they are horrified at ideas in whose formation they have not had a hand and in which they have not been able to purchase a substantial percentage of shares.

All bantering aside, let us now begin with a brief inventory of our ideas, at least those that seem most important. (Much of what follows was included in a talk presented by the author in Santiago, Chile, on May 23, 1991).

1. The Starting Point for Our Ideas

We do not initiate our conception of things with the affirmation of generalities, but rather in the study of the particulars of human life: what is particular to existence, what is particular to the personal register of thinking, feeling, and acting. This initial position means that the conception outlined here is incompatible with any system that starts from an *idea*, the *material*, the *unconscious*, the *will*, *society*, and so forth.

If someone accepts or rejects a given conception of things – however logical or eccentric it may be – it is always the person who is in play, accepting or rejecting this conception. The person does this, not society, or the unconscious, or matter.

Let us speak, then, of human life. When I observe myself, not from a physiological point of view but from an existential one, I find myself here, in a world that is given, neither made nor chosen by me. I find that I am in a situation with, in relationship with phenomena that, beginning with my own body, are inescapable. My body is at once the fundamental constituent of my existence and, at the same time, a phenomenon homogenous with the natural world in which it acts and on which the world acts. But the nature of my body has important differences for me from other phenomena, to wit:

1. I have an immediate register of my body;
2. I have a register, mediated by my body, of external phenomena; and
3. Some of my body's operations are accessible to my immediate intention.

2. The Human Being: Nature, Intention, and Opening

It happens, however, that the world appears not simply as a conglomeration of natural objects, it appears as an articulation of other human beings and of objects, signs, and codes they have produced or modified. The intention that I am aware of in myself appears as a fundamental element for the interpretation of the behavior of others, and just as I constitute the social world by comprehending intentions, so am I constituted by it.

Of course, I am speaking here of intentions that manifest in corporeal action. It is through the corporeal expressions of, or by perceiving the situation of the other, that I am able to comprehend the meanings of the other, the intention of the other. Moreover, natural or human objects appear as either pleasurable or painful to me, and so I modify my situation, trying to place myself in favorable relationship to them.

In this way, I am not closed to the world of the natural and other human beings, rather precisely what characterizes me is *opening*. My consciousness has been configured intersubjectively in that it uses codes of reasoning, emotional models, schemes of action that

I register as "mine," but that I also recognize in others. And, of course, my body is open to the world insofar as I both perceive and act over the world.

The natural world, as distinct from the human, appears to me as without intention. Of course, I can imagine that the stones, plants, and stars possess intention, but I find no way to hold an effective dialogue with them. Even those animals in which at times I glimpse the spark of intelligence appear as basically impenetrable to me and changing only slowly from within their own natures. I observe insect societies that are completely structured, higher mammals that employ rudimentary technology but still only replicate such codes in a slow process of genetic change, as if they were always the first representatives of their respective species. And when I observe the benefits of those plants and animals that have been modified and domesticated by humanity, I see human intention opening its way and humanizing the world.

3. The Human Being: Social and Historical Opening

To define human beings in terms of their sociability seems to be inadequate, because this does not distinguish them from many other species. Nor does capacity for work stand out as their most notable characteristic when compared to that of more powerful animals. Not even language defines them in their essence, for we know of numerous animals that use various codes and forms of communication.

All new human beings, in contrast, find themselves living in a world that is modified by others, and it is in their being constituted by this world of intentions that I discover their human capacity of accumulation within and incorporation to the temporal – that is, I discover not simply a social dimension but a socio-historical one.

Viewing things in this way, we can attempt a definition of the human being as follows: Human beings are historical beings whose mode of social action transforms their own nature. If I accept the above, I will also have to accept that such beings are capable of intentionally transforming their own physical constitutions. And this is just what is taking place.

This process began with the use of instruments by human beings which, placed before their bodies as external "prostheses," allowed them to extend the reach of their hands and their senses and to increase both their capacity for work and its quality. Although not endowed by nature to function in either aerial or aquatic environments, they have nevertheless created means to move through these media and have even begun to leave their natural environment, the planet Earth. Today, moreover, they have begun to penetrate their bodies, replacing organs, intervening in their brain chemistry, carrying out fertilization *in vitro*, and even manipulating their own genes.

If by the word "nature" one is trying to indicate something permanent and unchanging, then today this idea has been rendered seriously inadequate, even when applied to what is most object-like about human beings, that is, to their bodies. And in light of this, regarding any "natural morality," "natural law," or "natural institutions," it is clear that nothing in this field exists through nature, but on the contrary that everything is socio-historical.

4. The Transforming Action of the Human Being

Along with the conception of a human nature is another prevalent conception that has asserted the passivity of the consciousness. This ideology has considered the human being to be an entity that functions primarily in response to stimuli from the natural world. What began as crude sensualism has gradually been displaced by historicist currents that, at their core, have preserved the same conception of a passive consciousness. And even when they have emphasized the consciousness's activity in and transformation of the world more than the interpretation of its activities, they have still conceived of its activity as resulting from conditions external to the consciousness.

Today, these old prejudices regarding human nature and the

passivity of the consciousness are once again being asserted, this time transformed into neo-evolutionary theories embodying such views as natural selection, determined through the struggle for the survival of the fittest.

In the version currently in fashion, now transplanted into the human world, this sort of zoological conception attempts to go beyond earlier dialectics of race or class by asserting a dialectic in which it is supposed that all social activity regulates itself automatically according to "natural" economic laws. Thus, once again, the concrete human being is overwhelmed and objectified.

I have noted those conceptions that, to explain the human being, have begun from theoretical generalities and maintained the existence of an unchanging human nature and a passive consciousness. We maintain, quite the opposite, the need to start from human particularity, that the human being is a socio-historical and non-natural phenomenon, and that the human consciousness is active in transforming the world in accordance with its intention. We see human life as always taking place in a situation, and the human body as an immediately perceived natural object, immediately subject as well to numerous dictates of each person's intention. The following questions therefore arise:

- How is it that the consciousness is active, that is, how is it that its intentions can act upon the body and through the body transform the world?
- How is it that the human being is constituted as a socio-historical being?

These questions must be answered from particular existence so as not to fall again into theoretical generalities, from which a dubious system of interpretation might be derived.

To answer the first question, one must apprehend with immediate evidence how human intention acts over the body. To answer the second, one must begin from evidence of the temporality and intersubjectivity of the human being, rather than beginning from supposed general laws of history and society.

I will not go into greater detail here regarding these questions, as this would take us away from the broad themes of the present letter. For a more extensive treatment I refer you to two essays in the work *Contributions to Thought*[1] that deal with the above questions. The first essay, "Psychology of the Image," studies the function that the image fulfills in the consciousness, highlighting its aptitude for moving the body through space. The second essay, "Historiological Discussions," studies the theme of historicity and sociability.

5. Overcoming Pain and Suffering as Basic Vital Projects

In the work *Contributions to Thought* it is observed that the natural destiny of the human body is the world, and to verify this it is sufficient to observe the body's conformation. The body's sensory apparatus and those for feeding, locomotion, reproduction, and so on are naturally shaped to be in the world. Further, it is through the body that the image launches its transforming charge – not to copy the world, not to be a reflection of a given situation, but on the contrary to modify a given situation.

In the course of daily events, objects are either limitations on or amplifications of corporeal possibilities, and the bodies of others appear as a multiplication of those possibilities insofar as they are governed by intentions that are recognized as similar to those governing one's own body.

Due to the condition of finiteness and temporo-spatial limitation in which they find themselves and which they register as physical pain and mental suffering, human beings find it necessary to transform both the world and themselves. Overcoming pain is not simply an animal response, then, but a temporal configuration in which the future is paramount, and which becomes transformed into a fundamental impulse of life even though it may not be present as something urgent at any given moment. In this way, and aside from the immediate, reflex, and natural response to pain, the deferred response to avoid pain is spurred by psychological suffering in the face of danger, re-presented as future possibility or as present fact when pain is present in other human beings.

Overcoming pain, then, appears as a basic project that guides action. This is what has made possible communication among distinct human bodies and intentions in what is known as the *social constitution*. The social constitution is as historical as human life; it configures human life. Its transformation is ongoing, but in a different way than in nature where change does not occur as the result of intentions.

6. Image, Belief, Look, and Landscape

Let us suppose that one day I go into my room, and I see the window. I recognize it, it is familiar to me. I have not only a fresh perception of it, but also acting in me are my previous perceptions of it which, converted into images, have been retained within me. Suddenly, I notice a crack in one corner of the windowpane. "That wasn't there," I say to myself, on comparing the new perception with what I retain from my previous perceptions. And I also feel a sense of surprise.

The window of previous acts of perception has been retained in me, but not passively as in a photograph, rather actively, in the way that images function. What has been retained in me operates in the present with respect to what I perceive, even though the formation of those retentions pertains to the past. In this way the past is always present, always being updated.

Before entering my room I took it for granted, it was a given, that the window would be there in good condition. It was not that I was thinking about it, but simply that I was counting on it. The window itself was not explicitly present in my thoughts at that moment, rather it was co-present. It was within the horizon of objects contained in my room.

It is due to what is co-present, to this retention that is updated and superimposed on the perception, that the consciousness infers more than it perceives. And it is in this phenomenon that it is possible to see the most elemental functioning of belief. In this example I would say to myself: "I believed the window was in good condition."

If upon entering my room I had seen phenomena proper to a different field of objects, for example a motorboat or a camel, this surrealistic situation would have seemed unbelievable, not because those objects do not exist but simply because their location in my room would be outside the field of my co-presence, outside the landscape I have formed that acts within me, superimposing itself on every single thing that I perceive.

Now then, in any present instant of my consciousness I can observe the inter-crossing of what has been retained and what is being futurized in me as they act co-presently and in structure. In my consciousness, the present instant is constituted as an active temporal field of three different times. Here things take place very differently from the way they occur in calendar time, where today is separate and distinct from yesterday or tomorrow. On the calendar and on the clock, *now* is different from *no longer* and from *not yet*, and events are ordered one after the other in a linear succession that I cannot claim to be a structure, but is rather a subgroup within a complete series that I call a *calendar*. I will return to these ideas again when we consider the themes of historicity and temporality later on.

For now, let us continue with the previous notion that the consciousness infers more than it perceives, through its use of what comes from the past as retentions, superimposed on present perception. In each *look* or act of looking that I direct toward an object, what I see is distorted. This is not meant in the same sense that modern physics explains our inability to see the atom or wavelengths that lie above or below our thresholds of perception. What I am referring to is the distortion related to the superposition of the images of retentions and futurizations on perceptions in the present.

Thus, when I contemplate a beautiful sunset in the countryside, the natural landscape that I observe is not determined by and in itself. Rather, I determine it, I constitute it through the aesthetic ideal that I hold. And the special peace that I feel gives me the illusion that I contemplate passively, when in reality I am actively superimposing numerous of my own internal contents on the natural object itself. This phenomenon holds not only for the present example, but for all *looks* that I direct toward reality.

7. The Generations and Historical Moments

Social organization continues and expands, but this cannot take place solely through the presence of social objects that have been produced in the past, that we make use of in the present, and that we project into the future. Such a mechanism is too simple to explain the process of civilization.

Continuity is given by the different generations of human beings, which do not exist side by side, separate and apart from each other, but rather coexist, interact, and transform each other. These distinct generations, which make continuity and development possible, are dynamic structures. They are social time in motion, without which civilization would fall into a natural state, losing its character of being a society.

It happens, moreover, that in every historical moment the generations that coexist have distinct temporal levels, retentions, and futurizations that configure different landscapes of situation and belief. For the active generations, the bodies and behavior of children and the elderly constitute a presence that betrays where they have come from and where they are going. So, too, both ends of this triple relationship can recognize their extreme temporal positions. And this situation never stops or remains static, because while the active generations age and the elderly die, the children grow up and begin to occupy active positions. In the meantime, new births continuously reconstitute society.

When, as an abstraction, we "interrupt" this ceaseless flow, we can speak of a certain *historical moment* in which all members located in the same social setting can be considered *contemporaries*, living in one same time. But it should be noted that not all contemporaries are coetaneous, that is, they are not all the same age, nor do they have the same internal temporality in terms of landscapes of formation, present situation, and projects. What happens in fact is that a generational dialectic is established between those who are in the "layers" that lie closest to each other and who are trying to occupy the central activity, the social present, in accordance with their different interests and beliefs.

It is this internal social temporality, and not as some philosophies of history would have it the succession of phenomena placed linearly one after another as in calendar time, that structurally explains the historical becoming in which the different generational accumulations – that is, the accumulating landscapes of the distinct generations – interact.

Constituted socially in an historical world in which I continue to configure my landscape, I interpret that toward which I direct my *look*. This is my personal landscape, but it is in addition a collective landscape for large numbers of people in this time.

As has been previously observed, different generations coexist in the same present time. As an elementary example, those who were born before the transistor was invented and those born into the world of computers are both now living in the same moment. Numerous such coexisting configurations differ from each other in their experiences – not only in the ways that they act, but also in how they think and how they feel – and what used to work in one epoch regarding social relationships and modes of production has slowly, or at times abruptly, ceased to function.

People expected a certain result in the future; that future arrived, but things did not turn out as projected. And that former mode of action, that former sensibility, that former ideology – none of these any longer coincide with the new landscape now asserting itself in society.

8. Violence, the State, and the Concentration of Power

Human beings, through their opening, their freedom to choose between situations, their ability both to defer responses and to imagine their future, also have the possibility to negate themselves, to negate aspects of their bodies, to negate their bodies completely as in suicide, or to negate other human beings. It is this freedom that has allowed a few to illegitimately appropriate the social whole, that is, to negate the freedom and intentionality of others, reducing those others to prostheses, to instruments of the intentions of the few. Therein lies the essence of discrimination, with physical, economic, racial, sexual, religious and other forms violence as its methodology.

It is through power over the apparatus of social regulation and control, that is, the State, that violence can be established and perpetuated. Because of this, social organization will require an advanced type of coordination that is safe from any concentration of power, whether private or of the State.

When it is claimed that privatizing all areas of the economy will make society safe from the power of the State, what is not disclosed in this is that the real problem lies in the monopoly or oligopoly, which simply transfers power from the hands of the State to the hands of a para-State, no longer managed by a bureaucratic minority but now by that private minority itself as it continues to advance this process of concentration.

The various social structures from the most primitive to the most sophisticated are all proceeding toward ever greater concentration. Eventually they will reach the point that they become immobilized and begin a stage of dissolution, a stage that will give rise to new processes of reorganization, but at a higher level than before.

From the beginning of history, society has proceeded toward globalization, and there will come a time of maximum concentration of arbitrary power, displaying the character of a world empire, which will be without any further possibilities of expansion. The collapse of this global system will follow the logic of the structural dynamics of all closed systems, in which disorder necessarily tends to increase.

Just as the process of the current structures tends toward globalization, however, so does the process of humanization proceed toward increasing opening of the human being, moving beyond both the State and para-State toward decentralization and de-concentration in favor of a superior form of coordination among autonomous social particularities.

Whether everything ends up in chaos and civilization starts anew, or we begin a stage of progressive humanization, does not depend on inexorable mechanical designs, but on the intentions of individuals and peoples, on their commitment to changing the world, and on an ethic of liberty, which by definition is something that cannot be imposed. And we will aspire no longer to formal democracy, controlled until now by the special interests of the various factions, but instead to real democracy in which direct participation can be realized instantaneously, thanks to communication technologies that are every day more able to bring this about.

9. The Human Process

Those who have diminished the humanity of others have in so doing necessarily brought about new pain and suffering, rekindling in the heart of society the age-old struggle against natural adversity – but now between on one side those who wish to "naturalize" other human beings, society, and history, and on the other side the oppressed, who need to humanize themselves in humanizing the world. That is why to humanize is to move beyond objectification to affirm the intentionality of every human being and the primacy of the future over the present situation.

It is the image and representation of a future that is both better and possible that allows the modification of the present and makes every revolution and all change possible. This is why the pressure of oppressive conditions is not in itself sufficient to set change in motion, rather it is necessary to realize that such change is possible and that it depends on human actions.

This struggle is not between mechanical forces, it is not a natural reflex. It is, rather, a struggle between human intentions. And that is precisely what permits us to speak of oppressors and the oppressed, of the just and the unjust, of heroes and cowards. This is the only thing that allows the meaningful practice of social solidarity and commitment to the liberation of those who suffer discrimination, whether they are a majority or a minority.

For more detailed considerations regarding violence, the State, institutions, the law, and religion, and so as not to exceed the limits of this brief letter, I refer you to the work entitled *The Human Landscape*.[2]

I do not believe that the meaning of human actions has to do with senseless upheavals or "useless passions" that end in nothing but absurd disintegration. I believe that the destiny of humanity is oriented by intention, and that as people become increasingly conscious of this intention it opens the way toward a *universal human nation.*

From what we have previously seen **it is abundantly clear that human existence does not simply begin and end in a vicious circle of self-enclosure, and that a life aspiring to coherence must open itself, expanding its influence toward people and social ambits, advancing not only a concept or a few ideas but precise actions that extend the growth of freedom.**

In the next letter I will leave aside these strictly doctrinal themes in order to focus once more on themes involving the current situation and personal action in the social world.

With this letter I send my

Warmest regards,
Silo

Fifth Letter to My Friends

June 4, 1992

Dear Friends,

Along with many people who are concerned about the unfolding of present-day events, I frequently find myself in the company of those who have been active in progressive political parties and organizations. Many of them have yet to recover from the shock they received with the fall of "real socialism." Today, all over the world, activists by the hundreds of thousands are choosing to withdraw into the concerns of their daily lives, making it understood with this attitude that they believe their old ideals have been foreclosed. What for me represented simply one more episode in the disintegration of centralized structures – indeed something anticipated for over two decades – came for them as an unexpected catastrophe.

Yet this is not the time for everyone to simply drop out of sight, because as the current political form dissolves this leaves a disparity of forces that is opening the way for a system that is monstrous in both its conduct and its direction.

A couple of years ago I attended a rally where older workers, working mothers with their children, and small groups of young people raised their clenched fists together as they sang the words to their anthem in unison. Their banners were waving as the echoes of their glorious calls to struggle rolled across the scene. And upon seeing this I thought of just how much good will, risk, tragedy, and striving, all moved by heartfelt convictions, had been lost along a road leading to the absurd negation of any possibilities of transformation.

How much I would have liked to accompany that moving scene with a song to the ideals of old militants – those who, giving no thought to the outcome, remained steadfast in their combative pride. And all of this gave rise in me to strongly mixed feelings, and today at a distance I ask myself: What has happened to the many good people who struggled in solidarity for something greater than their own immediate interests, for what they believed would be the best of worlds? I am thinking not only of those who were members of more or less institutionalized political parties, but of all those who chose to place their lives at the service of a cause they believed was just. And, of course, one cannot take their measure solely by cataloguing their errors or by classifying them as the exponents of a particular political philosophy. Today it is imperative to redeem human courage, inspiring people's ideals in a new and possible direction.

In reading over the first part of this letter, I must apologize to those who, not having participated in those movements and activities, may feel removed from such themes. At the same time I would point out to them the importance of keeping these matters in mind – matters that bear so directly on the values and ideals of human actions. These, then, are the themes with which today's letter deals, perhaps a bit firmly, but with the intention of shaking off the defeatism that seems to have taken such deep hold in the militant soul.

1. The Most Important Issue: To Know If One Wants to Live, and In What Conditions

Today, millions of people struggle simply to subsist, not knowing whether tomorrow they will be able to surmount hunger, disease, and neglect. Their needs are so dire that whatever they undertake to escape their problems only further complicates their lives. Are they to do nothing then, and remain in a state that is really only one of postponed suicide? Are they to attempt desperate measures?

What sort of activity, what risk, what prospect are they prepared to face? What are those, who for economic, societal, or simply personal reasons find themselves in extreme situations, supposed to do? Always, the most important question is to know if one wants to live, and to decide in what conditions to do so.

2. Human Liberty: Source of All Meaning

Even those who do not find themselves in extreme situations are today questioning whether their present circumstances can form a way of life in the future. Even those who prefer not to think about their situation, or who turn this responsibility over to others, are still choosing a way of life. Thus, freedom of choice is a reality from the moment we question our lives and reflect on the conditions in which we want to live. Whether we then struggle for that future or not, this freedom of choice still exists. And it is only this fact of human life that can justify the existence of values, of morality, of law, and of obligation, just as it also allows us to refute all politics, all forms of social organization, and every way of life that is imposed without justifying its meaning, without substantiating just how it is at the service of the concrete human being in today's world. Any morality, any law, or any social constitution that begins from principles supposedly superior to human life places that life in a situation of contingency, denying its essential meaning of liberty.

3. Intention: Orientor of Action

We are born into conditions that we have not chosen. We have not chosen our body, our natural environment, our society, or the space and time we have either the luck or the misfortune to occupy. Subsequently, there is a point at which we acquire the liberty to commit suicide or to go on living and to reflect on the conditions in which we want to live. We can rebel against a tyranny and be victorious or die in the attempt; we can struggle for a cause or facilitate oppression; we can accept a model of life or try to change it. And we can also make a mistake in our choice.

We may believe that by accepting everything that is established in a society, no matter how perverse those things are, we are becoming more perfectly adapted, and this is the path to better conditions in our lives. Or instead, we may think that by questioning everything, without distinguishing between what is of primary importance and what is secondary, we will expand the range of our liberty – when in reality our power to change things diminishes in a phenomenon in which we become increasingly less adapted. Finally, we can give priority to actions that extend our influence in a new direction, one that is possible for us, one that gives meaning to our existence. In every case, we will have to choose among conditions, among needs, and we will do so according to our intention and the vision of life that we propose for ourselves. Of course, our intention itself can continue to change along this path that is so subject to accidents.

4. What Should We Do with Our Lives?

We cannot ask ourselves this question in the abstract, but only in relation to the concrete situation in which we live and the conditions in which we wish to live. For now, we exist in a particular society and in relationship with other people, and our destinies are interwoven with their destinies. If we believe that at present everything is fine and what we can glimpse of the future seems satisfactory for us as individuals and for society, then we need only forge ahead, perhaps with some minor reforms, but certainly in the same direction. If, on the contrary, we think that we live in a violent, unjust society that is filled with inequity and assailed by unremitting crises related to the dizzying changes in the world, then we will reflect at once on the need for profound personal and social transformations.

Affected by the global crisis now sweeping us along, we lose stable references, and planning our futures becomes ever more difficult. More serious still is our inability to carry out coherent action to change this situation, both because the familiar forms of struggle have failed and also because the unraveling of the social fabric makes it increasingly difficult to mobilize significant numbers of people.

Of course, the same thing happens to us that happens to everyone who is experiencing the present difficulties and intuitively grasps just how much conditions are deteriorating. No one can or would want to undertake actions that are destined to fail, and yet no one can simply let things go on this way.

And the worst of it is that by our inaction we open the door to even greater inequity and injustice. Forms of discrimination and abuse long thought overcome are resurfacing with greater virulence than ever. Given such disorientation and crisis, what is to prevent new monstrosities from acting as social references, forms whose representatives will not only state but also enforce what each and every one of us is to do? Such primitive occurrences are becoming more possible than ever because today their simplistic message spreads so easily, reaching those who find themselves in extreme situations.

More and more people, whether well or poorly informed, have come to recognize that by now we are in a situation of crisis that can be characterized in approximately the terms used here. Nevertheless, the option they are following with increasing single-mindedness is to focus only on their own lives, ignoring the difficulties of others and everything that is taking place in the social context around them.

Many times, while we applaud the objections that others make against the prevailing system, we ourselves are very far from trying to do anything that could actually change those conditions. We know that today democracy is merely formal, responding as it does to the dictates of the economic interests. Yet, subject to the blackmail of either supporting that system or facilitating the rise of dictatorships, we salve our consciences with ridiculous votes for major parties.

It is not reasonable to believe that the act of voting for and asking others to vote for small parties can constitute a phenomenon of interest in the future, nor will support for forming labor organizations outside the established frameworks be an important factor in bringing people together.

And because we view such work as too limited, we reject those efforts that are rooted in neighborhoods, in communities, in urban areas, and in our immediate environments. It is clear, however, that this is where the rebuilding of the social fabric will begin when the crisis finally overtakes the centralized structures.

Yet instead of keeping our ears attuned to the undercurrent of the people's demands for change, we prefer to focus on the superficial game of the powerful elites, the famous, the formers of opinion. We object to the actions of the mass media controlled by economic interests, instead of dedicating ourselves to exercising influence in the smaller media and taking advantage of the many openings for social communication. And if we continue to work as militants within some progressive political organization, our usual tactic is to try to dredge up some incoherent character who can get us "press," some famous personality who can represent our current of thought because he is more or less palatable to the news media of the prevailing system.

Basically, all of this happens to us because we believe we are defeated and that we have no other recourse than to nurse our growing bitterness in silence. And we call this defeat "dedicating ourselves to our own lives." Meanwhile, "our own lives" accumulate contradictions as we lose touch with the meaning of and any capacity to choose the conditions in which we want to live. Eventually, we cannot even conceive of the possibility of a great *movement for change* that can serve as a reference, drawing together the most positive factors in society. And of course our previous disappointments keep us from acting as protagonists in this process of transformation.

5. Moral Consciousness and Short-Term Interests

We have to choose the conditions in which we want to live. If we go against our life project we will not escape from contradiction, which will leave us at the mercy of a long chain of accidents. And if we take that direction, what brake can we then apply to slow the cascading events of our lives? Only that of our short-term interests. In our resulting lives of expediency, then, we can imagine extreme situations of every kind befalling us, from which in our rush to escape we will sacrifice every value and all meaning, because our sole focus has become our own immediate benefit.

To avoid such difficulties, we shun any commitment that could draw us toward extreme situations, but of course events themselves will necessarily put us in positions that we have not chosen. It does not require any special brilliance to understand what is sure to happen with those closest to us should they adopt this same position – if they pursue identical benefits, will they not then be in opposition to us? And what is to prevent our whole society from following this same path? In this situation of arbitrariness without limits, naked power will overwhelm everything before it. Where it encounters resistance it will do so with overt violence. Where it doesn't, it will make do with persuasion that relies for justification on untenable values, to which we will all have to submit even while in the depths of our hearts we experience how meaningless life has become. And if this comes to pass it will mark the triumph of the Earth's dehumanization.

To choose a life project within imposed conditions is far from being a simple animal reflex. On the contrary, it is the essential characteristic of the human being. And if we eliminate this quality – which defines the human being – we block human history, and we can expect only the advance of destruction at every step. If we give up the right to choose a life project and an ideal of society, we will find ourselves left with only caricatures of law, values, and meaning. Under such circumstances, what will we then uphold in the face of the neurosis and upheaval we are beginning to experience all around us?

Each of us will have to see what to do with his or her own life, but all of us will have to bear in mind as well that our actions extend beyond ourselves, and this is so regardless of whether our capacity to influence others is great or small. The choice between unifying actions – those with meaning – or contradictory actions dictated by immediacy, is inescapable in every situation in which the direction of life is at issue.

6. Sacrificing One's Objectives for Circumstantial Success: Some Habitual Errors

Everyone who is committed to collective action, every person who works with others toward meaningful social objectives, needs to be clear on the numerous errors that have in the past brought ruin upon the best of causes. Ridiculous Machiavellian schemes, personality clashes placed above mutually agreed upon goals, and authoritarian behavior of every stripe fill volumes of history books, as well as our personal memories.

By what right does anyone use a doctrine, a plan of action, a human organization, only to push aside the priorities they themselves have expressed? What right do we have to propose to others an objective and a destiny, only to later place as the primary value some supposed success or need of the moment? What would be the difference between this and the pragmatism we say we repudiate? In following that path, how could there be any coherence among what one thinks, feels, and does?

In every age, "instrumentalists" have committed the same moral fraud of presenting others with an inspiring image of the future, gaining for themselves an immediate image of success. In then sacrificing the intention agreed upon, however, they open the door to

negotiating every sort of betrayal with the faction against which they claim to struggle. And this indecency is then justified by some supposed "need" concealed within the initial proposal.

It should be clear that I am not speaking of those changes of conditions and tactics in which all involved understand the connection to the agreed-upon objectives that mobilized them in the first place. Nor am I referring to those mistakes in evaluating situations that can occur in the process of carrying out concrete actions. These observations apply to the immorality that distorts intentions and against which it is indispensable to be alert. It is important to be attentive to ourselves as activists, and also to explain this to others so they understand beforehand that if they break their commitments this will leave our hands as free as theirs.

There is, of course, a whole range of clever tricks for using other people, and there is no way to catalogue them all. Nor will we become "moral censors," because it is clear that behind this attitude lies a repressive form of consciousness. The objective of such people is to sabotage any action they do not control, immobilizing their companions in struggle with mutual mistrust. And when they smuggle in as contraband from another field supposed values by which they judge our actions, it is good to remember that it is their "morality" that is in question, and that it is not the same as ours. Why, then, would such people choose to be with us?

Finally, it is important to be aware of a less-than-honest gradualism that is used to manipulate situations until in reality they come to oppose their stated objectives. It is in this position that we find all those who accompany us with motives different from those they express. Their mental direction is twisted from the beginning and awaits only the opportunity to manifest itself. In the meantime, they gradually expand their use of codes that, whether overtly or covertly, embody a system of double-speak. This attitude is almost always found among those people who, in the name of some militant organization, disorient activists of good faith, while at the same time they endeavor to make responsibility for their abuses fall on the shoulders of authentic militants.

It is not my intention here to dwell on the familiar "internal problems" that affect every human organization, but it does seem useful to mention the opportunistic root that underlies this behavior, which involves introducing a mobilizing image of the future, gaining for oneself an immediate image of success.

7. The Kingdom of the Secondary

Present circumstances are such that accusers of every stripe and description adopt a prosecutorial tone and demand explanations from us, acting as though it is we who must prove our innocence to them. What is noteworthy is that their basic tactic lies in exalting all that is secondary, and as a consequence obscuring the primary questions.

This attitude recalls the practice of democracy within companies. Employees may discuss, for example, whether the desks in the office should be nearer to or farther from the windows and whether the office should be furnished with flowers or painted in pleasant colors, none of which is in itself bad. Then they vote, and the majority decides the fate of the furniture and the color of the paint, and this is also not in itself bad. But when it comes time to discuss and propose taking a vote on questions of management and operation, a terrified silence falls...instantly any idea of democracy is frozen, because in reality we are dwelling in the *kingdom of the secondary*.

Nothing different can be expected from the "prosecutors" of the system. Suddenly some journalist will take on that role – making a preference some of us may have for certain types of food, for example, seem somehow suspicious, or demanding that we "take a stand" on today's burning questions of sports, astrology, and the catechism. Of course, they are never lacking for some clumsy accusation, to which it is assumed we must respond, and in superficially setting the context they bandy about words charged with double meanings as they manipulate contradictory images.

What is important to remember is that those who choose to

locate themselves in a faction opposed to us have every right to have us explain to them why they are in no condition to judge us and why we, on the other hand, are fully justified in judging them. They need to realize that it is they who must defend their position against our objections. Of course, whether this can actually take place in any given instance depends on certain conditions being present and the individual skill of the contenders, but it is always most exasperating to see people who have every right to take the initiative bow their heads before such incoherence.

It is pathetic, too, to watch various leaders on the television screen as they mouth their witticisms and dance like trained bears with the host of the program, or to see them submitting to every sort of humiliation just to make the front pages. Yet, as they watch these wonderful examples, many well-intentioned people fail to realize the extent to which the message they are viewing has been deformed or diluted by the time the mass media release it to the public at large.

These comments have focused on facets of the kingdom of the secondary that operate by displacing attention from the fundamental issues, with the result that what reaches the public – supposedly to enlighten them – is really disinformation. Curiously, a great many progressive people are taken in by this trap, failing to understand very clearly just how their receiving this abundance of apparent "news" in practice leaves them more bewildered than accurately informed.

Finally, this is no time to let languish in the camp of the opposition positions that in reality we need to defend. Were we to abandon these positions, anyone could reduce our position to mere frivolity simply by affirming that he, too, is for example a "humanist" because he is concerned about what is human; that he is "nonviolent" because he deplores war; that he is against discrimination because he has a black friend or a communist friend; that he is an environmentalist because he agrees that we need to protect seals and trees. If pressed, however, such people will be incapable of backing up in any depth the superficial things they say – and the mask will slip, showing their real face, which is anti-humanist, violent, discriminatory, and predatory.

While the previous commentaries on these expressions of the kingdom of the secondary do not really contribute anything new, it is nonetheless worthwhile from time to time to alert those naive activists who, in trying to communicate their ideas, have yet to realize just how strange is this kingdom of the secondary in which they have been interned.

I hope that you will be able to overlook any discomfort experienced on reading a letter perhaps so little related to your own problems and interests, and I trust that in the next letter we will be able to go on to more pleasant things.

With this letter I send my

Warmest regards,
Silo

Sixth Letter to My Friends

April 5, 1993

Dear Friends,

In further correspondence, certain readers of my letters have continued their critiques, demanding greater definition of social and political action as well as the prospects for such efforts to transform the present state of affairs. In these circumstances I could simply confine myself to restating what I wrote at the beginning of the first letter: "For some time now I have been receiving correspondence from various countries requesting that I explain or elaborate on certain of the subjects addressed in my books. For the most part what they have sought are explanations about such concrete issues as violence, politics, the economy, the environment, as well as social and interpersonal relationships. As you can see, these concerns are many and varied, and it is clear that the answers will have to come from specialists in these fields, which of course I am not."

Although I have offered commentaries on these topics in subsequent letters, it seems that I have not yet managed to satisfy their requests. And this leaves me with a difficulty, for how am I to respond to questions of such broad scope in a writing the length and nature of a letter?

As you know, I participate in a current of opinion, in a movement that during three decades of activity has given rise to numerous institutions and has confronted dictatorships and injustices of every stripe. The efforts of those in this movement have been met with a varied mixture of disinformation, defamation, and deliberate silence. Yet despite these difficulties, this movement has spread around the world, while preserving both its financial and its ideological independence. Had it yielded to expediency, engaging in the usual sordid short-term speculation, it would doubtless have received recognition and press. But this would only have finally consecrated the triumph of the absurd and the victory of everything against which it has struggled.

In its history, the blood of those who participate in this movement has been shed. They have faced imprisonment, deportation, and barriers of every kind. And it is necessary to remember this. In this sense our movement has always felt a close kinship as a tributary of historical humanism, which placed such emphasis on freedom of conscience in the struggle against all obscurantism and in defense of the highest human values. But our movement has also produced works and studies sufficient to provide responses for this era, in which events have finally precipitated a profound crisis. And I will appeal to these works and studies in order to set forth, within the limits of this letter, the fundamental themes and proposals of contemporary humanists.

DOCUMENT OF THE HUMANIST MOVEMENT

Humanists are women and men of this century, of this time. They recognize the achievements of humanism throughout history, and find inspiration in the contributions of many cultures, not only those that today occupy center stage. They are also men and women who recognize that this century and this millennium are drawing to a close, and their project is a new world. Humanists feel that their history is very long and that their future will be even longer. As optimists who believe in freedom and social progress, they fix their gaze on the future, while striving to overcome the general crisis of today.

Humanists are internationalists, aspiring to a *universal human nation.* While understanding the world they live in as a single whole, humanists act in their immediate environments. Humanists seek not a uniform world, but a world of multiplicity: diverse in ethnicity, languages and customs; diverse in local and regional autonomy; diverse in ideas and aspirations; diverse in beliefs, whether atheist or religious; diverse in occupations and in creativity.

Humanists do not want masters, they have no fondness for authority figures or bosses. Nor do they see themselves as representatives or bosses of anyone else. Humanists want neither a centralized state nor a para-state in its place. They want neither armed gangs nor a police state in their place.

But a wall has arisen between humanist aspirations and the realities of today's world. The time has come to tear down that wall. To do this, all humanists of the world must unite.

I. GLOBAL CAPITAL

This is the great universal truth: Money is everything. Money is government, money is law, money is power. Money is basically sustenance, but more than this it is art, it is philosophy, it is religion.

Nothing is done without money, nothing is possible without money. There are no personal relationships without money, there is no intimacy without money. Even peaceful solitude depends on money.

But our relationship with this "universal truth" is contradictory. Most people do not like this state of affairs. And so we find ourselves subject to the tyranny of money – a tyranny that is not abstract, for it has a name, representatives, agents, and well-established procedures.

Today, we are no longer dealing with feudal economies, national industries, or even regional interests. Today, the question is how the surviving economic forms will accommodate to the new dictates of international finance capital. Nothing escapes, as capital worldwide continues to concentrate in ever fewer hands – until even the nation state depends for its survival on credit and loans. All must beg for investment and provide guarantees that give the banking system the ultimate say in decisions. The time is fast approaching when even companies themselves, when every rural area as well as every city, will all be the undisputed property of the banking system. The time of the para-state is coming, a time in which the old order will be swept away.

At the same time, the traditional bonds of solidarity that once joined people together are fast dissolving. We are witnessing the disintegration of the social fabric, and in its place find millions of isolated human beings living disconnected lives, indifferent to each other despite their common suffering. Big capital dominates not only our objectivity, through its control of the means of production, but also our subjectivity, through its control of the means of communication and information.

Under these conditions, those who control capital have the power and technology to do as they please with both our material and our human resources. They deplete irreplaceable natural resources and act with growing disregard for the human being. And just as they have drained everything from companies, industries, and whole governments, so have they deprived even science of its meaning – reducing it to technologies used to generate poverty, destruction, and unemployment.

Humanists do not overstate their case when they contend that the world is now technologically capable of swiftly resolving the problems in employment, food, health care, housing, and education that exist today across vast regions of the planet. If this possibility is not being realized, it is simply because it is prevented by the monstrous speculation of big capital.

By now big capital has exhausted the stage of market economies, and has begun to discipline society to accept the chaos it has itself produced. Yet in the presence of this growing irrationality, it is not the voices of reason that we hear raised in dialectical opposition. Rather, it is the darkest forms of racism, fundamentalism, and fanaticism that are on the rise. And if groups and whole regions are increasingly guided by this new irrationalism, then the space for constructive action by progressive forces will diminish day by day.

On the other hand, millions of working people have already come to recognize that the centralized state is as much a sham as capitalist democracy. And just as working people are standing up against corrupt union bosses, more than ever citizens are questioning their governments and political parties. But it is necessary to give a constructive orientation to these phenomena, which will otherwise stagnate and remain nothing more than spontaneous protests that lead nowhere. For something new to happen, a dialogue about the fundamental factors of our economy must begin in the heart of the community.

For humanists, labor and capital are the principal factors in economic production, while speculation and usury are extraneous. In the present economic circumstances, humanists struggle to totally transform the absurd relationship that has existed between these factors. Until now we have been told that capital receives the profits while workers receive wages, an inequity that has always been justified by the "risk" that capital assumes in investing – as though working people do not risk both their present and their future amid the uncertainties of unemployment and economic crisis.

Another factor in play is management and decision making in the operation of each company. Earnings not set aside for reinvestment in the enterprise, not used for expansion or diversification, are increasingly diverted into financial speculation, as are profits not used to create new sources of work.

The struggle of working people must therefore be to require maximum productive return from capital. But this cannot happen unless management and directorships are cooperatively shared. How else will it be possible to avoid massive layoffs, business closures, and even the loss of entire industries? For the greatest harm comes from under-investment, fraudulent bankruptcies, forced acquisition of debt, and capital flight – not from profits realized through increased productivity. And if some persist in calling for workers to take possession of the means of production following nineteenth-century teachings, they will have to seriously consider the recent failures of real socialism.

As for the argument that treating capital the same way work is treated will only speed its flight to more advantageous areas, it must be pointed out that this cannot go on much longer because the irrationality of the present economic system is leading to saturation and crisis worldwide. Moreover, this argument, apart from embracing a radical immorality, ignores the historical process in which capital is steadily being transferred to the banking system. As a result, employers and business people are being reduced to the status of employees, stripped of decision-making power in a lengthening chain of command in which they maintain only the appearance of autonomy. And as the recession continues to deepen, these same business people will begin to consider these points more seriously.

Humanists feel the need to act not only on employment issues, but also politically to prevent the State from being solely an instrument of international capital, to ensure a just relationship among the factors of production, and to restore to society its stolen autonomy.

II. REAL DEMOCRACY VERSUS FORMAL DEMOCRACY

The edifice of democracy has fallen into ruin as its foundations – the separation of powers, representative government, and respect for minorities – have fallen into ruin.

The theoretical separation of powers is nonsense. Even a cursory examination of the practices surrounding the origin and composition of the different powers reveals the intimate relationships that link them to each other. And things could hardly be otherwise, for they all form part of one same system. In nation after nation we see one branch gaining supremacy over the others, functions being usurped, corruption and irregularities surfacing – all corresponding to the changing global economic and political situation of each country.

As for representative government, since the extension of universal suffrage people have believed that only a single act is involved when they elect their representative and their representative carries out the mandate received. But as time has passed, people have come to see clearly that there are in fact two acts: a first in which the many elect the few, and a second in which those few betray the many, representing interests alien to the mandate they received. And this corruption is fed within the political parties, now reduced to little more than a handful of leaders who are totally out of touch with the needs of the people. Through the party machinery, powerful interests finance candidates and then dictate the policies they must follow. This state of affairs reveals a profound crisis in the contemporary conception and implementation of representative democracy.

Humanists struggle to transform the practice of representative government, giving the highest priority to consulting the people directly through referenda, plebiscites, and direct election of candidates. However, in many countries there are still laws that subordinate independent candidates to political parties, or rather to political maneuvering and financial restrictions that prevent them from even reaching the ballot and the free expression of the will of the people.

Every constitution or law that prevents the full possibility of every citizen to elect and to be elected makes a mockery of real democracy, which is above all such legal restrictions. And in order for there to be true equality of opportunity, during elections the news media must be placed at the service of the people, providing all candidates with exactly the same opportunities to communicate with the people.

To address the problem that elected officials regularly fail to carry out their campaign promises, there is also a need to enact *laws of political responsibility* that will subject such officials to censure, revocation of powers, recall from office, and loss of immunity. The current alternative, under which parties or individuals who do not fulfill their campaign promises risk defeat in future elections, in practice does not hinder in the least the politicians' second act – betraying the people they represent.

As for directly consulting the people on the most urgent issues, every day the possibilities to do so increase through the use of technology. This does not mean simply giving greater importance to easily manipulated opinion polls and surveys. What it does mean is to facilitate real participation and direct voting by means of today's advanced computational and communications technologies.

In real democracy, all minorities must be provided with the protections that correspond to their right to representation, as well as all measures needed to advance in practice their full inclusion, participation, and development.

Today, minorities the world over who are the targets of xenophobia and discrimination make anguished pleas for recognition. It is the responsibility of humanists everywhere to bring this issue to the fore, leading the struggle to overcome such neo-fascism, whether overt or covert. In short, to struggle for the rights of minorities is to struggle for the rights of all human beings.

Under the coercion of centralized states – today no more than the unfeeling instruments of big capital – many countries with diverse populations subject entire provinces, regions, or autonomous groups to this same kind of discrimination. This must end through the adoption of federal forms of organization, through which real political power will return to the hands of these historical and cultural entities.

In sum, to give highest priority to the issues of capital and labor, real democracy, and decentralization of the apparatus of the State, is to set the political struggle on the path toward creating a new kind of society – a flexible society constantly changing in harmony with the changing needs of the people, who are now suffocated more each day by their dependence on an inhuman system.

III. THE HUMANIST POSITION

Humanist action does not draw its inspiration from imaginative theories about God, nature, society, or history. Rather, it begins with life's necessities, which consist most elementally of avoiding pain and moving toward pleasure. Yet human life entails the additional need to foresee future necessities, based on past experience and the intention to improve the present situation.

Human experience is not simply the product of natural physiological accumulation or selection, as happens in all species. It is social experience and personal experience directed toward overcoming pain in the present and avoiding it in the future. Human work, accumulated in the productions of society, is passed on and transformed from one generation to the next in a continuous struggle to improve the existing or natural conditions, even those of the human body itself. Human beings must therefore be defined as historical beings whose mode of social behavior is capable of transforming both the world and their own nature.

Each time that individuals or human groups violently impose themselves on others, they succeed in detaining history, turning their victims into "natural" objects. Nature does not have intentions, and thus to negate the freedom and intentions of others is to convert them into natural objects without intentions, objects to be used.

Human progress in its slow ascent now needs to transform both nature and society, eliminating the violent animal appropriation of some human beings by others. When this happens, we will pass from pre-history into a fully human history. In the meantime, we can begin with no other central value than the human being, fully realized and completely free. Humanists therefore declare, "Nothing above the human being, and no human being beneath any other."

If God, the State, money, or any other entity is placed as the central value, this subordinates the human being and creates the condition for the subsequent control or sacrifice of other human beings. Humanists have this point very clear. Whether atheists or religious, humanists do not start with their atheism or their faith as the basis for their view of the world and their actions. They start with the human being and the immediate needs of human beings. And if, in their struggle for a better world, they believe they discover an intention that moves history in a progressive direction, they place this faith or this discovery at the service of the human being.

Humanists address the fundamental problem: to know if one wants to live, and to decide on the conditions in which to do so.

All forms of violence – physical, economic, racial, religious, sexual, ideological, and others – that have been used to block human progress are repugnant to humanists. For humanists, every form of discrimination, whether subtle or overt, is something to be denounced.

Humanists are not violent, but above all they are not cowards, and because their actions have meaning they are unafraid of facing violence. Humanists connect their personal lives with the life of society. They do not pose such false dichotomies as viewing their own lives as separate from the lives of those around them, and in this lies their coherence.

These issues, then, mark a clear dividing line between humanism and anti-humanism: humanism puts labor before big capital, real democracy before formal democracy, decentralization before centralization, anti-discrimination before discrimination, freedom before oppression, and meaning in life before resignation, complicity, and the absurd. Because humanism is based on freedom of choice, it offers the only valid ethic of the present time. And because humanism believes in intention and freedom, it distinguishes between error and bad faith, between one who is mistaken and one who is a traitor.

IV. FROM NAIVE HUMANISM TO CONSCIOUS HUMANISM

It is at the base of society, in the places where people work and where they live that humanism must convert what are now only simple isolated protests into a conscious force oriented toward transforming the economic structures.

The struggles of spirited activists in labor unions and progressive political parties will become more coherent as they transform the leadership of these entities, giving their organizations a new orientation that, above short-range grievances, gives the highest priority to the basic proposals advocated by humanism.

Vast numbers of students and teachers, already sensitive to injustice, are becoming conscious of their will to change as the general crisis touches them. And certainly, members of the press in contact with so much daily tragedy are today in favorable positions to act in a humanist direction, as are those intellectuals whose creations are at odds with the standards promoted by this inhuman system.

In the face of so much human suffering, many positions and organizations today encourage people to unselfishly help the dispossessed and those who suffer discrimination. Associations, volunteer groups, and large numbers of individuals are on occasion moved to make positive contributions. Without doubt, one of their contributions is to generate condemnations of these wrongs. However, such groups do not focus their actions on transforming the underlying structures that give rise to the problems. Their approaches are more closely related to humanitarianism than to conscious humanism, although among these efforts are many conscientious protests and actions that can be extended and deepened.

V. THE ANTI-HUMANIST CAMP

As the people continue to be suffocated by the forces of big capital, incoherent proposals arise that gain strength by exploiting people's discontent, focusing it on various scapegoats. At the root of all such neo-fascism is a profound negation of human values. Similarly, there are certain deviant environmental currents that view nature as more important than human beings. No longer do they preach that an environmental catastrophe is a disaster because it endangers humanity – instead to them the only problem is that human beings have damaged nature.

According to certain of these theories, the human being is somehow contaminated, and thus contaminates nature. It would have been better, they contend, had medicine never succeeded in its fight against disease or in prolonging human life. "Earth first!" some cry hysterically, recalling Nazi slogans. It is but a short step from this position to begin discriminating against cultures seen to contaminate or against "impure" foreigners. These currents of thought may be considered anti-humanist because at bottom they hold the human being in contempt, and in keeping with the nihilistic and suicidal tendencies so fashionable today, their mentors reflect this self-hatred.

There is, however, a significant segment of society made up of perceptive people who consider themselves environmentalists because they understand the gravity of the abuses that environmentalism exposes and condemns. And if this environmentalism attains the humanist character that corresponds, it will direct the struggle against those who are actually generating the catastrophes – big capital and its chain of destructive industries and businesses, so closely intertwined with the military-industrial complex.

Before worrying about seals they will concern themselves with overcoming hunger, overcrowding, infant mortality, disease, and the lack of even minimal standards of housing and sanitation in many parts of the world. They will focus on the unemployment, exploitation, racism, discrimination, and intolerance in a world that is so technologically advanced, yet still generates serious environmental imbalances in the name of ever more irrational growth.

One need not look far to see how the right wing functions as a political instrument of anti-humanism. Dishonesty and bad faith reach such extremes that some exponents periodically present themselves as representatives of "humanism." Take, for example, those cunning clerics who claim to theorize on the basis of a ridiculous "theocentric humanism." These people, who invented religious wars and inquisitions, who put to death the very founders of western humanism, are now attempting to appropriate the virtues of their victims. They have recently gone so far as to "forgive the errors" of those historical humanists, and so shameless is their semantic banditry that these representatives of anti-humanism even try to cloak themselves with the term "humanist."

It would of course be impossible to list the full range of resources, tools, instruments, forms, and expressions that anti-humanism has at its disposal. But having shed light on some of their more deceptive practices should help unsuspecting humanists and those newly realizing they are humanists as they re-think their ideas and the significance of their social practice.

VI. HUMANIST ACTION FRONTS

With the intention of becoming a broad-based social movement, the vital force of humanism is organizing action fronts in the workplace, neighborhoods, unions, and among social action, political, environmental, and cultural organizations. Such collective action makes it possible for varied progressive forces, groups, and individuals to have greater presence and influence, without losing their own identities or special characteristics. The objective of this movement is to promote a union of forces increasingly able to influence broad strata of the population, orienting the current social transformation.

Humanists are neither naive nor enamored of declarations that belong to more romantic eras, and in this sense they do not view their proposals as the most advanced expression of social consciousness or think of their organization in an unquestioning way. Nor do they claim to represent the majority. Humanists simply act according to their best judgment, focusing on the changes they believe are most suitable and possible for these times in which they happen to live.

◊ ◊ ◊

This Humanist Document gives greater definition of certain aspects of contemporary humanism, and in the next letter we will go on to consider other matters.

With this letter I send my

Warmest regards,
Silo

Seventh Letter to My Friends

August 7, 1993

Dear Friends,

This letter will speak of social revolution. But how is this possible, since certain arbiters of opinion have already explained that following the collapse of real socialism the word *revolution* has fallen out of fashion? Perhaps in the back of their minds is the belief that all revolutions prior to 1917 were simply precursors to the "real" revolution. And if the real revolution has failed, clearly this is a subject that may no longer be discussed.

As is their custom, these right-thinking people continue to exercise ideological censorship, assuming the prerogative of conferring or denying legitimacy on words and fashions. The views of these bureaucrats of the spirit (or more precisely, of the media) continue to be diametrically opposed to ours: Previously such people believed the Soviet monolith to be eternal, while today they view the triumph of capitalism as an unalterable reality. They take it for granted that the substance of any revolution must involve bloodshed, accompanied by an indispensable backdrop of marches, gestures, fiery speeches, and banners waving in the breeze.

Hollywood cinematography and Pierre Cardin fashions were constantly present in their formative landscapes, so that today when they consider Islam, for example, they think of women's dress, which causes them much concern. And when they speak of Japan, as soon as they have discussed the economic plan they can hardly wait to express their indignation that the kimono has never quite been phased out. If as children they were raised on a diet of books and movies about pirates, later they felt drawn to Katmandu, island vacations, preserving the environment, and "natural" fashions. If instead they relished westerns and action movies, later they viewed progress in terms of a war of competition and revolution in terms of gunpowder.

We are immersed in a world of codes of mass communication in which the formers of public opinion impose their message through newspapers, magazines, radio, and television; a world in which writers of limited intelligence determine which themes may even be discussed; a world in which reasonable people inform us about today's events and explain to us the way things work. The company of those who may express opinion gather daily before the cameras. There in civilized fashion the psychologist, the sociologist, the political consultant, the fashion expert, the journalist who interviewed Khadafy, and the ineffable astrologer hold forth, one after another. And then all of them shout at us in unison: "Revolution? But that's so completely *passé*!" In short, public opinion (that is, published opinion) maintains that everything is improving, despite a few setbacks, and certifies, moreover, the demise of the revolution.

But what body of well-articulated ideas has been presented to discredit the revolutionary process in today's world? To date nothing more serious than talk-show opinions. In the absence of vigorous conceptions that merit rigorous discussion, let us go on at once to matters of substance.

1. Destructive Chaos or Revolution

This series of letters presents a number of commentaries regarding the general situation in which we now live. These descriptions lead to the following dilemma: Either we let ourselves be swept along by the tendency toward a world that is ever more absurd and destructive, or we give events a different direction. Underlying this formulation is the dialectic of freedom versus determinism, the human search for choice and commitment versus the acceptance of mechanical tendencies and processes with their dehumanizing end.

The continuing concentration of big capital to the point of worldwide collapse would be dehumanizing, as would be the results: a world convulsed by hunger and overflowing with refugees; a world of endless fighting, warfare, chaos, and constant fear; a world of abuse of authority, injustice, and erosion of basic liberties; a world in which new forms of obscurantism will triumph. It would be dehumanizing to go once more round the same circle until some other civilization arises, only to mechanically repeat the same stupid steps again – that is, if this is still possible after the collapse of the first planetary civilization that is now beginning to take shape.

Within this long history, however, one's own life and the life of each generation is so short and so immediate that **one sees the wider destiny of all as a simple extension of one's own destiny, rather than one's own destiny as a particular case of the wider destiny.**

So it is that the lives people live today are far more compelling than any thought of the life that they or their children will live tomorrow. And, of course, for millions of human beings the situation is so urgent that they have no horizon left to consider some hypothetical future that might come to pass.

At this very moment there are already far too many tragedies, and this is more than enough reason to struggle for a profound change in the overall situation. Why, then, do we speak of tomorrow, if the pressing problems of today are so great? Simply, because the image of the future is increasingly manipulated and we are admonished to put up with present circumstances as if this crisis were something insignificant to bear. "Every economic adjustment," their theories go, "has a social cost." "It is regrettable," we are told, "that for all of us to be well off in the future you will have to endure these hard times today." "And when before," they ask, "has there ever been such technology and medical care as the wealthy nations have today?" "Soon," they assure us, "your time will come, too."

And while they put us off, the actions of those who promise

progress for all continue to widen the gap that separates the opulent few from the majorities who suffer ever-greater outrages. The prevailing social order locks things into a vicious circle feeding on itself as it expands into a worldwide system from which no part of the planet is free.

It is also clear, however, that as positions become more radical and unrest grows more widespread, people everywhere are beginning to see through the hollow promises of society's leaders.

Will everything end up, then, in the war of all against all? Will the future be culture against culture, continent against continent, region against region? Will it be ethnic group against ethnic group, neighbor against neighbor, and family member against family member as people flail about without direction like wounded animals trying to shake off their pain? Or instead will we include and welcome all the differences within the direction of world revolution?

What I am trying to express is that **we are facing the alternative of either destructive chaos, or revolution as a direction that goes beyond the differences among those who are oppressed.** I am saying that each day both the global situation and the particular situation of each individual will become more filled with conflict, and it would clearly be suicidal to leave our future in the hands of the same people who have directed this process so far.

No longer do we live in times in which one can simply wipe out all opposition and then the following day proclaim, "Peace reigns in Warsaw." These are not times in which ten percent of the population can do as they please with the other ninety percent.

Yet today the world is becoming a single closed system where, in the absence of a clear direction for change, capital and power simply continue to accumulate at the expense of everything else. The result is that within this closed system one can expect nothing more than a continued mechanical increase in general disorder. And the paradox of closed systems tells us that any attempt to impose order on the growing disorder will only further accelerate the growth of that disorder. The only way out of this predicament is to revolutionize the system, opening it up to the diversity of human needs and aspirations. Proposed in these terms, the theme of revolution takes on more than usual importance, with a scope and ramifications it could not have had in former times.

2. Of What Revolution Are We Speaking?

The previous letter outlined positions regarding the questions of labor versus big capital, real democracy versus formal democracy, decentralization versus centralization, anti-discrimination versus discrimination, and freedom versus oppression.

If at present capital is steadily being transferred to the banking system, if the banking system continues to gain ownership of companies, nations, regions, and the world, then revolution implies that the banking system be transformed so that its services are made available without charging usurious interest.

If a company is constituted so that capital receives the profits while the workers receive salaries or wages, if company management and decision making rest solely in the hands of capital, then revolution implies that profits be reinvested, diversified, or used to create new sources of employment, and that management and decision making be shared by labor and capital.

If the regions, provinces, or states within a country have their hands tied by centralized decision making, then revolution implies restructuring that centralized power into regional entities forming a federal republic, and for those regions to be similarly decentralized in favor of locally based power, from which all electoral representation must derive.

If health and education are provided in an unequal way to the inhabitants of a country, then revolution implies free access to education and health care for everyone, because these are clearly the two highest values of the revolution and must replace wealth and power in the current social paradigm. **Viewing everything in terms of the priorities of education and health care provides the correct framework for dealing with the highly complex economic and**

technological challenges facing today's society. It seems that in no other way, certainly not while wealth and power remain the highest values, can a society with evolutionary possibilities be formed.

The central argument employed by capitalism against new proposals is to cast doubt on them by continually asking where the financial resources will come from and how productivity will be increased, implying by this that it is only lending by the banking system and not the work of the people that is the origin of resources. Besides, what is the purpose of productivity if this production simply vanishes at once from the hands of those who produce it?

Nor are we taught anything extraordinary by the model of society that has been in place for some decades in certain parts of the world (and that is now beginning to fall apart). Whether education and health care are really progressing so remarkably in those countries still remains to be seen in light of the growing plagues, which are not only physical but also psycho-social.

If it is part of education to create an authoritarian, violent, and xenophobic human being, if part of progress in health care is rising alcoholism, drug addiction, and suicide, then such a model is obviously not valid. Although as humanists we will continue to admire the well-organized centers of education and the well-equipped hospitals, we will endeavor to ensure that **they are placed at the service of all people without distinction.** However, in regard to the content and meaning of education and healthcare, there are more than ample grounds on which to object to the present system.

This letter speaks of a social revolution that will result in a dramatic change in people's living conditions, of a political revolution that will alter the power structure, and ultimately of a human revolution that will create its own paradigms in replacing today's decadent values. **The social revolution to which humanism aspires will come to pass through gaining the political power necessary to carry out appropriate transformations, but gaining that power is not in itself an objective.** Moreover, violence is not an essential component of this revolution. What good would it be to follow the repugnant practices of imprisoning and executing one's enemies? What would be the difference between this and what oppressors have always done?

India's anti-colonial revolution was brought about by popular pressure and not through violence, and while this revolution remained unfinished due to the limited scope of its ideology, it did demonstrate a new methodology of action and struggle. The revolution that overthrew the Iranian monarchy was also unleashed by popular pressure; a takeover of the centers of political power was not even necessary as these were already "emptied," destructured, until eventually they ceased to function altogether. Then the intolerance that followed ruined everything.

Thus, revolutions are possible by various means, including electoral victory. But in every case drastic transformations of society's structures must immediately be set in motion – beginning with the establishment of a new legal order that, among other things, will fully exhibit the new social relationships of production, prevent abuses of power, and modify the function of those structures that, although they come from the past, are still capable of being improved.

Today, however, neither the revolutions that are dying nor the new ones being born will progress past the stage of speeches within this stagnating social order. They will not develop beyond the stage of organized mobs if they do not advance in the direction signaled by humanism, that is, toward a system of social relationships whose central value is the human being, and not other values such as "productivity" or "a socialist society," for example.

But to place the human being as the central value implies an idea that is totally distinct from what is generally understood today by the term *human being*. The current models used to characterize the human being are still far removed from the idea and the sensibility necessary to fully grasp the reality of what is human. Still, and it is important to point this out, beyond the confines of today's naive and superficial models there are some signs of a revival of critical intelligence. To mention but one case, the work of G. Petrovich[3] embodies concepts that presage the present development. He defines revolution as "the creation of an essentially distinct mode of being, different from all being that is non-human, anti-human, and not-yet-entirely-human." Petrovich concludes by identifying revolution with the highest form of being, as "being in fullness" and "Being-in-Liberty."

The revolutionary tide already in motion expresses the desperation of the oppressed majorities, and it will not be stopped. But this alone will not be enough, because a suitable direction for this process will not come about solely through the mechanisms of "social practice." **What is imperative at this time, when the human being is so completely circumscribed, is to move from the field of necessity to the field of liberty by means of revolution. Future revolutions, if they are to be more than putsches, palace coups, or the simple redress of class, ethnic, or religious grievances, will have to take on an inclusive and transforming character based on what is essentially human. And beyond the changes they will produce in the concrete situations of their countries, their character will be universalist and their objectives globalizing. Thus, when we speak of "world revolution" it is understood that the character and objectives of any humanist revolution or any revolution that becomes humanist, though it may take place in a limited area, will carry it beyond itself. And every such revolution, no matter how insignificant the location in which it takes place, will involve the essentiality of every human being.** World revolution cannot simply be proposed in terms of "success," but rather in its real and humanizing dimension. Moreover, the new kind of revolutionary who corresponds to this new type of revolution becomes, by essence and by activity, a humanizer of the world.

3. Action Fronts in the Revolutionary Process

Next I would like to expand on certain practical considerations related to creating the conditions necessary for a social force of sufficient unity, organization, and growth to position itself in the direction of a revolutionary process.

Today, the old thesis of forming common fronts among progressive forces based on minimum points of agreement has in practice become only "clusters" of partisan dissidents clinging together without connection to the wider society. The result is that contradictions accumulate among their leaders, who are reduced merely to pursuing media coverage and political self-promotion. During times when a well-funded political party could achieve hegemony over many fragments, it was viable to propose forming common fronts for electoral campaigns. Today, despite the fact that the situation has changed drastically, the traditional left continues to follow these same procedures as if nothing were different.

It is necessary to review the function of the political party in today's world and to ask whether parties are structures that are still capable of setting revolution in motion. For if the prevailing system has completed the process of swallowing political parties, reducing them to hollow shells in an artificial activity controlled by big capital and the banking system, then a party of mere superstructure without any human base could achieve formal power (but not real power) without in the process necessarily introducing even minimal fundamental change.

For now, political action requires creating a party that attains electoral representation at various levels. It must be clear from the outset, however, that the objective of such representation is to direct the conflict to the heart of established power. In that context, a party member who becomes a representative of the people is not so much a public functionary as a reference who calls attention to the contradictions of the system, organizing the struggle in the direction of the revolution. In other words, party or institutional political work is understood here as the expression of a broader social phenomenon that has its own dynamic. In this way, while a party may reach its greatest level of activity during elections, the different action fronts that from time to time form its base will use these same elections to call public attention to social conflicts and to broaden their organizations.

Here we find important differences from the traditional conception of a party. Indeed, until only a few decades ago the party was thought of as the vanguard of the struggle, bringing together different action fronts. The proposal here is just the opposite: Action fronts organize and develop the base of a social movement, while a party becomes the institutional expression of this movement. In turn, such a party must create conditions so that other progressive political forces will be fully included; it cannot expect them to lose their identity and simply blend in. This party must reach beyond its own identity and form a broad-based front with other forces to include the many progressive factors that are now so fragmented. But this will amount to nothing more than agreements among leadership unless the party has a real base that orients the process.

This proposal is not, however, reversible; that is, this party cannot form part of a front organized by other entities that are merely superstructures. Such a party, whose real strength comes from the base organization, can form a political front with other forces that agree with certain basic conditions established by this party.

Let us now consider the various types of action fronts. Such fronts need to work in the administrative base of each country, focusing on city and local government. The idea is to develop in the workplaces and neighborhoods of the selected areas **common fronts committed to actions that address real conflicts that have been correctly prioritized. This last point means that working to redress short-term grievances is meaningless if that struggle does not result in organizational growth and positioning for subsequent steps.** It is important to make it clear to everyone just how each conflict is directly related to their standard of living, to health care, and to education (and as their understanding deepens, workers in the fields of health and education will tend to become direct supporters and later form part of the cadres necessary for directly organizing the social base).

The same phenomena that we find taking place with political parties in the present system are also occurring in unions and labor organizations. Thus, the proposal is not to win control of labor organizations or unions but to bring together the workers who will as a consequence replace the former leadership's control. In this area it is important to encourage all systems of direct elections as well as any conventions and assemblies that commit the leadership – requiring either that they take positions on concrete conflicts that provide meaningful responses to the demands of the base or be bypassed. And certainly, labor action fronts must design their tactics with the objective of growth in the organization of the social base.

Finally, setting in motion social and cultural institutions that act from the base is of the utmost importance, because it allows communities that suffer discrimination or persecution to come together in a context of respect for human rights, finding a common direction notwithstanding their particular differences. The thesis that all ethnic groups, collectivities, and human groupings subject to discrimination must become strong by themselves so as to confront the abuse they are subject to exhibits a significant lack of understanding of the predicament we are all in. It is a position that stems from the notion that "mixing" with foreign elements will cause a loss of identity, when in reality it is precisely their isolated position that leaves them exposed and easily eradicated, or else left in a situation where they become so radical that their persecutors can justify direct action against them. **The best guarantee of survival for minorities suffering discrimination is for them to form part of an action front with others to channel the struggle for their demands in a revolutionary direction.** After all, it is the system taken as a whole that has created the conditions for discrimination, and these conditions will not disappear until that social order is transformed.

4. Revolutionary Process and Revolutionary Direction

It is important to distinguish between revolutionary process and revolutionary direction. From our point of view, **a revolutionary process is understood as a set of mechanical conditions that are generated as the system develops.** In this sense, such development creates factors of disorder that are ultimately either supplanted, assert themselves, or end up causing a breakdown of the entire scheme of things. According to this analysis, the globalization toward which the world is now proceeding is generating acute factors of disorder in the overall development of the system. And, as we have discussed in previous letters on more than one occasion, this process is independent of the voluntary action of groups or individuals. The problem that now arises is what, precisely, will be the future of this system, given that it is mechanically proceeding to revolutionize itself without the intervention of any progressive orientation whatsoever.

The orientation at issue depends on human intention and escapes the determinism of the conditions produced by the present system. I have already presented on previous occasions my position on the non-passivity of the human consciousness, its essential quality of not being simply a reflection of objective conditions, its capacity to oppose such conditions and to devise a future situation different from life at present. (See "Fourth Letter to My Friends," sections 3 and 4, pp. 43–46, and *Contributions to Thought* [4]).

It is within this mode of liberty, within conditions, that we interpret the revolutionary direction.

It is through the exercise of violence that a minority of the wealthy and powerful impose their conditions on the social whole, organizing an order – an inertial system – that simply continues its mechanical development. Viewed in this way, the modes of production as well as the resulting social relationships, the legal order, the dominant ideologies that regulate and justify this order, and the apparatus of the State or para-State by means of which the whole of society is controlled, are all revealed as instruments that serve the interests and intentions of the minority holding power. But the system continues to develop mechanically beyond the intentions of the powerful few as they endeavor to concentrate ever more the factors of power and control, in the process only further accelerating the process of the prevailing system, which increasingly escapes their control.

The resulting disorder will clash with the established order, provoking the powers that be to apply proportionately greater resources for their protection. In critical periods, the whole of society will be disciplined with all the violence that the system has at its disposal. And this leads to the maximum recourse available: the armed forces. Is it entirely certain, however, that the armed forces will continue to respond in the traditional way during times such as these when the whole system is heading toward a global collapse? If they do not, the momentous shift in the direction of current events that could result is a subject that merits further discussion.

Even a brief examination of the final stages of the civilizations that have preceded the present one shows that armies have indeed risen up against the established powers, and have as well become divided by the civil wars for which the seeds were already present. But because the system was unable by itself to introduce a new direction into this situation, it simply proceeded along its catastrophic course. Will the world civilization now taking shape suffer the same fate? In the next letter we will have to further consider the case of the armed forces.

With this letter I send my

Warmest regards,
Silo

Eighth Letter to My Friends

August 10, 1993

Dear Friends,

As indicated in the previous letter, the present letter will focus on points related to the armed forces. The interest of this writing will center, of course, on the relationships among the armed forces, political power, and society, and will be based on the paper I presented three months ago in Moscow under the title "The Need for a Humanist Position in the Contemporary Armed Forces."[5] This letter will depart from the concepts in the original paper in treating the position of the military in the revolutionary process, a topic that will allow further development of ideas outlined in previous letters.

1. The Need to Redefine the Role of the Armed Forces

Today the armed forces are endeavoring to define what their new role will be in a process that began with the proposals for progressive proportional disarmament initiated by the Soviet Union toward the end of the 1980s. The diminishing tensions between the superpowers led to a reversal in the concept of defense for the major powers. Meanwhile, the gradual replacement of military-political blocs, in particular the Warsaw Pact, by a system of relatively cooperative relationships has unleashed centrifugal forces that have given rise to fresh conflicts in various parts of the globe. Certainly, at the height of the cold war limited wars were frequent and often prolonged, but the current character of these conflicts has changed, and now threatens to spill over from the Balkans into the Muslim world and other areas of Asia and Africa.

Given the secessionist tendencies inside many countries, the border disputes that previously occupied the armed forces of adjoining nations are today taking a different direction. Economic, ethnic, and linguistic differences are leading to changes in borders long thought unalterable at the same time that large-scale migrations are taking place. Human groups are being uprooted as they flee desperate situations; others try to hold back or expel different groups from certain areas.

These and other phenomena reveal profound changes, particularly in the structure and conception of the State. At the same time that we are witnessing a process of economic and political regionalization, we are also seeing growing discord within many countries as they move toward this regionalization. **It is as if the nation state, designed two hundred years ago, is no longer able to withstand the blows from above by multinational interests and from below by the forces of secession.** Increasingly dependent, increasingly tied to the regional economy, increasingly pitted against other regions in economic warfare, the State is undergoing a crisis of unprecedented proportions as it struggles to maintain control of the changing situation in which it finds itself.

Existing civil and commercial laws and regulations have become obsolete, and constitutional documents are being amended to open the way for the ever-greater worldwide movement of capital and financial resources. Even penal codes are changing – today a citizen may be seized for a crime that has been tried in another country under foreign laws by judges of a different nationality. The traditional concept of national sovereignty, then, has been noticeably weakened. The entire legal-political apparatus of the State, its institutions, and those people directly or indirectly in its service, are all experiencing the effects of this general crisis.

The armed forces, long assigned the role of protector of the general sovereignty and security, are also suffering these problems. As education, health care, and the means of communication are privatized along with goods and services, natural resources, and even significant areas of public safety, this continues to erode the importance of the traditional State. It follows that if the administration and resources of a nation are removed from the sphere of public control, that the legal and judicial system will follow suit, reducing the armed forces to the role of a mere private militia assigned to defending only parochial or multinational financial interests. And indeed these trends have recently intensified in many countries.

2. Continuing Factors of Aggression in This Period of Reduced Tensions

Among the powers that have declared the cold war at an end, external aggression has yet to disappear, however. Violations of air and maritime space continue, as do provocations against distant nations, fresh incursions and base installations, new military pacts, and even foreign wars and occupations to control shipping lanes or areas with abundant natural resources.

The clear record established in the wars in Korea, Vietnam, Laos, and Cambodia; in the Suez, Berlin, and Cuban crises; and in the invasions of Grenada, Tripoli, and Panama, has shown the world that more powerful nations frequently direct disproportionate military action against defenseless nations, a record that weighs heavily at the time of disarmament talks.

This type of action is particularly grave when, as in the case of the Gulf War, it takes place on the flanks of important powers that could interpret such acts as threats to their own security. These excesses also have harmful secondary effects when they strengthen certain sectors within those powers, allowing them to criticize their governments as incapable of stopping such encroachments. And all of this could compromise the climate of international peace that is now so vital.

3. Internal Security and Military Restructuring

Regarding internal security, it is important to note two problems that are visible on the horizon: social explosions and terrorism.

If unemployment and recession continue to rise in the industrialized nations, it is possible these areas will be the scene of growing unrest and upheavals, reversing to a degree the picture of previous decades in which conflicts arose on the periphery of these centers, which were nonetheless able to continue their expansion without experiencing undue shocks. Today, however, events such as the recent riots in Los Angeles could spread beyond one city and even to other countries.

Secondly, the phenomenon of terrorism presents a danger of some magnitude, considering the firepower to which these relatively specialized individuals and groups now have access. This threat could take the form of high explosives and even nuclear devices or chemical and biological weapons, all of which continue to become less expensive and easier to produce.

In the unstable panorama of today's world, the concerns of the armed forces are many and varied. In addition to the strategic and political problems they face, there are internal issues of restructuring, large-scale troop reductions, recruiting and training methods, replacement of equipment, technological modernization and, of primary importance, declining budgets. However, while the armed forces must thoroughly comprehend these factors in the context of their own sphere of activity, it must be added that none of these problems can be fully resolved until the primary function that the military is to fulfill in society and the world is made clear. It is, after all, political power that gives orientation to the armed forces, which must act in accordance with that orientation.

4. A Review of the Concepts of Sovereignty and Security

In the traditional conception of these issues, the armed forces are assigned the function of safeguarding the nation's sovereignty and security and granted the authority to use force in accordance with the mandate of the duly constituted powers. In this way, the State's monopoly on violence is transferred to the military services.

But this brings us to a key point in the discussion of what should be understood by the terms *sovereignty* and *security*. If a nation's sovereignty and security or, in more modern terms, its "progress" are said to require extraterritorial sources of raw materials, indisputable rights of maritime passage to protect the flow of commerce, and the control of strategic points or the occupation of foreign territory with these same objectives, then what we are faced with is the theory and practice of colonialism or neocolonialism.

The function of the military during colonial times consisted principally of facilitating the interests of the crowns of the period, and later on the interests of the private companies that obtained special concessions of political power in exchange for suitable compensation. The illegality of that system was justified by the supposed barbarism of the subjugated peoples, who were characterized as incapable of adequately governing themselves. The ideology corresponding to this stage affirmed colonialism as a "civilizing" system *par excellence*.

During the age of Napoleonic imperialism, the function of the army, which also held political power, consisted of expanding the borders with the declared objective of redeeming through military action peoples who were oppressed by tyrannies, and installing a legal and administrative system enshrining liberty, equality, and fraternity in its legal codes. The corresponding ideology justified this imperial expansion by the claim of "necessity" on the part of a power constituted by the democratic revolution against illegal monarchies that were based on inequality and that moreover formed a united front to suppress the revolution.

More recently, and following the teachings of Clausewitz, war has been understood as a simple extension of politics, and the State as promoter of these policies is considered the governing apparatus of a society that lies within certain geographical limits. Starting with this premise, geo-politicians have reached certain definitions they now hold dear in which borders are viewed as the "skin of the State," and in this organic-logical conception, the "skin" contracts or expands in accordance with the vital energy of the nation, and must thus expand as the progress of the community demands greater "living space" given its population density or economic strength.

From this perspective, the function of the military is to acquire space according to the demands of the policy of security and sovereignty, which is given primacy over the needs of neighboring countries. In this case, the dominant ideology proclaims that the differences in the needs experienced by various collectivities are related to "inherent" characteristics. This zoological vision of the struggle for the survival of the fittest recalls Darwinian conceptions, here illegitimately carried into the sphere of political and military practice.

5. The Legality and Limits of Established Power

Today we frequently hear reference made to the three conceptions used above to illustrate both how the military responds to political power and how it is framed within the various positions that political powers define at any given time as security and sovereignty. And if the function of the military is to serve the State in matters of security and sovereignty, and if the conception of these two factors varies from government to government, then the armed forces will have to abide by these changing directions.

Are there any limits or exceptions to this? **Two clear exceptions can be seen: 1) when political power has been illegally constituted and civil recourse to rectify this irregular situation has been exhausted; and 2) when political power has been legally constituted but in its exercise has become illegal, and civil recourse to rectify this anomalous situation has been exhausted.**

In both cases the armed forces have the duty to reestablish the legality that has been interrupted, which is equivalent to carrying forward the actions civil means were unable to bring about. In such

circumstances the military's duty is clearly to the law and not to the established power.

This does not mean, however, promoting a state that is dependent on the military; rather, the focus is on restoring the legality previously interrupted by an established power of criminal origin or one that has become criminal.

The questions that must now be asked are: where does legality originate, and what are its characteristics? As humanists our view is that legality flows from the people, as it is the people who give rise to a particular type of State and fundamental laws, to which the citizenry must then submit. And in the extreme case that the people should decide to amend the type of State and type of laws, it is incumbent upon the State and the legal system to carry this out, because the existence of no state structure or legal system can be placed above such a decision by the people. This point leads to a consideration of the revolutionary act, which will be treated further on.

6. Military Responsibility to Political Power

It should be emphasized that the military services need to be made up of citizens who recognize and carry out their responsibilities to the legality of the established power. And if the established power is functioning based on a democracy in which the will of the majority is respected through the election and replacement of representatives of the people, in which minorities are respected in accordance with established law, and in which there is respect for the separation and independence of powers, then the armed forces need not pass judgment on the correctness or errors of their government.

If, however, an illegal regime is imposed, then the armed forces cannot simply support it by mechanically invoking "obligatory obedience" to this regime. And in the case of international conflicts, the armed forces cannot carry out genocide following the orders of a political power made feverish by abnormal circumstances. For if human rights are not placed above every other right, it is not possible to understand why either social organization or the State exist. In the same way, no one can claim to be "just following orders" when it comes to assassination, torture, or degradation of the human being. If the trials following World War II taught us anything, it is that every person in the military has responsibilities as a human being, even in the extreme situation of armed conflict.

At this point it could be asked: Is not the military an institution whose training, discipline, and equipment make it primarily a factor of destruction? I would reply that long ago things were set up as they are today and, independent of the aversion we feel for every form of violence, we cannot now propose the simple disappearance or unilateral disarmament of the military, which would only leave a vacuum that will be filled by other aggressive forces, as was previously noted in relation to the record of attacks carried out against defenseless nations.

The armed forces have an important mission to fulfill in not obstructing the philosophy and practice of proportional progressive disarmament and through inspiring their colleagues in other countries to move in the same direction. They can make it clear that the function of the military in the world today is to avoid both catastrophes and the oppression imposed by illegal governments that do not answer to a popular mandate.

The greatest service, then, that the armed forces can contribute to their country and to all of humanity will be to prevent the existence of war. This proposal, which might seem utopian, is today supported by the strength of events that demonstrate how dangerous and impractical it is for everyone when military power increases, either unilaterally or globally.

Let us now return to the theme of military responsibility through some examples of the opposite. During the period of the cold war the West repeated two messages: that NATO and other alliances were formed to preserve a way of life threatened by Soviet and on occasion Chinese communism, and that military actions were undertaken in distant lands to protect the "interests" of the Western powers.

In Latin America the military preferred the pretext of the threat of internal subversion to justify their *coups d'état*. The armed forces there failed to answer to political power, trampling all law and every constitution in militarizing practically an entire continent under this so-called "doctrine of national security." The sequel of death and backwardness left by these dictatorships was bizarrely justified throughout the chain of command by the concept of "obligatory obedience," holding that under military discipline each level is simply to follow the orders of the next higher level. This way of posing things, reminiscent of Nazi justifications of genocide, must be borne in mind in any discussion of the limits of military discipline.

Our point of view, as already mentioned, is that once the military severs its dependence on political power it then constitutes an irregular force, an armed gang outside the law. This issue is clear, but admits one exception: a military uprising against a political power that has been illegally established or subsequently become seditious. The armed forces cannot invoke "obligatory obedience" to such an illegal power or they become accomplices in this irregularity, just as in other circumstances they cannot engage in a military coup, ignoring their duty to follow the popular mandate. These issues relate to internal order and, similarly, during international conflicts the armed forces cannot attack the civilian population of an enemy nation.

7. Military Restructuring

Regarding military recruitment, our point of view favors replacing compulsory with optional military service, a system that allows superior training of the professional soldier. And this limitation on personnel level also coincides with an interest in significant reduction in the levels of enlisted and officer personnel.

It is clear that a satisfactory restructuring of the military cannot be accomplished without attending to the personal, family, and social problems that will arise in numerous armies that now find themselves oversized for today's world. The new employment positions, geographical locations, and re-entry into society of these troops will be more harmonious if the military maintains a flexible relationship with them throughout the period required for their readjustment.

The primary factor that must be taken into account in the restructuring taking place today in various parts of the world is the political model of each country involved. Naturally, a unitary political system has characteristics different from those of a federal system or one in which various countries are joining together to form a regional community.

Our point of view favors federal systems open to regional confederations, for which a correctly designed restructuring will require permanent solid commitments to give continuity to this project. Without a clearly established desire on the part of all the parties to move in this direction, such restructuring will not be possible, because the financial support from each country will be subject to unpredictable political fluctuations. In this case, the federal armed forces would have only a formal existence, and military contingents would be the simple aggregation of separate troops from each community that is part of the federation. Attempting to form a unified command in this situation will present serious problems that will be difficult to resolve. In short, the political power that orients the military must set the guidelines, and in each set of circumstances the armed forces will require very precise and coordinated guidance.

Another important problem in restructuring is related to security forces. The function of security forces, if they are not part of the military, is to maintain internal order and to protect the country's citizens, although habitually they become involved in operations of surveillance and control of the population that are far removed from the objectives for which they were created. In many countries, the organizational chart in which they appear shows them directly connected to political ministries or cabinets of the interior rather than the ministry of war or defense.

The police, in contrast, are understood as public servants formed to follow a legal chain of command that will not be detrimental to the country's inhabitants; they have an accessory character and fall under the jurisdiction of the judicial branch. Often, however, through their character as a public force they carry out operations that can make them appear like military forces in the eyes of the population. It is clear how inappropriate such confusion is, and that it is in the best interest of the armed forces that these distinctions be made clear to all.

Similar things occur with other State organizations such as the intelligence services or other secret bodies, which often overlap and duplicate each other and which also have nothing to do with the military. The military does need an appropriate system of gathering intelligence to operate efficiently, but not one that in any way resembles mechanisms of surveillance and control of the country's citizens, because the military's function has to do with the security of the nation and certainly not with involvement in any ideological approval or censure by the government of the moment.

8. The Military's Position in the Revolutionary Process

It is supposed that in a democracy power flows from the sovereignty of the people. Both the conformation of the State as well as of those organizations that derive from it stem from this same source. Thus, in defending the country's sovereignty and providing security for the country's inhabitants, the military fulfills the function conferred on it by the State.

As we have seen, aberrations can of course occur if the military or some other faction illegally seizes power. And as we have also seen, the extreme case can occur in which the people may decide to change the type of State and type of laws – that is, the entire system. Under these circumstances, it is incumbent on everyone to carry out these changes, because there is no state structure or legal system whose existence can be placed above this decision by the people.

Certainly the fundamental documents of many countries contemplate the possibility that these documents can be modified by popular decision. This is one way that revolutionary change can take place, through which formal democracy will give way to real democracy.

If, however, this possibility is blocked, that constitutes a denial of the source from which all legality flows. In these circumstances, and only after having exhausted all civil recourse, it is the obligation of the military to carry out this will to change by removing the faction that is currently installed, now illegally, in power over public life. Through that military intervention society can reach the creation of revolutionary conditions in which the people can put into practice a new type of social organization and a new legal system.

It is hardly necessary to point out the differences between military intervention having the objective of returning to the people the sovereignty that has been stolen from them and the simple military coup that violates the legality previously established by popular mandate. Consistent with these ideas, legality requires that the will of the people be respected even when they propose revolutionary changes. Why shouldn't the majority express their desire to change these basic structures and, what is more, why shouldn't minorities have the opportunity to work politically to bring about revolutionary change in society? **Denying the will to revolutionary change through repression and violence seriously compromises the legality of the current system of today's formal democracies.**

It will be observed that this letter has not touched on matters relating to military strategy or doctrine or on questions of military technology and organization. This could not have been otherwise, for **we have applied a humanist point of view to the armed forces in relation to political power and society.**

The men and women of the military still have before them the enormous task, both theoretical and practical, of adapting their framework and organization to this special time in which we find our world. The views of society, and the genuine interest of the armed forces, although not specialists, to know those views, are a matter of fundamental importance. At the same time, vigorous relationships between members of the military from different countries, accompanied by frank and civilized discussions, form important steps toward recognizing the plurality of points of view. The attitude in some military forces that keeps them isolated from others, and their unresponsiveness in the face of the people's demands, correspond to an earlier period in which human and tangible interchange were restricted. Today the world has changed for everyone, including the armed forces.

9. Considerations on the Military and Revolution

Two widely asserted opinions are today of special interest: The first declares that the time of revolutions has passed and the second that military influence in decision making is gradually declining. It is also supposed that only in certain backward or poorly organized countries do such hindrances from the past still pose a threat. It is further held that as the system of international relations takes on an ever more solid character, it will make its weight felt until all the old factors of disorder are brought under control.

On the question of revolutions, as already noted, our point of view is diametrically opposed to the above notions. Whether concerted action by "civilized" nations will impose a new world order in which military influence will play no part is highly debatable. It should be noted that it is precisely in those nations and regions that are taking on an imperial character that both revolutions and military influence are increasingly making their presence felt. Sooner or later, as the forces of money become ever more concentrated, they will confront the majority, and in this situation *bank* and *military* will end up being antithetical terms.

As contemporary humanists we find ourselves, then, at the antipodes of the interpretation of historical processes from supporters of the prevailing system. Only the times near at hand will tell which perception of events is correct, events that some always seem to find (in the tradition of recent years) "incredible." With their way of looking at things, what will they say when the things described here do come to pass? Probably that humanity has gone backwards, returned to the past, or in more everyday terms that "the world has fallen apart."

We believe that phenomena such as the spread of irrationality, the rise of ever stronger religiosity, and many other related phenomena do not belong to the past, but correspond to a new stage that we will have to face with all the intellectual courage and human commitment of which we are capable. It will not work to go on claiming that society can best develop by staying the present course. What is important here is to comprehend that the conditions under which we are living are leading us directly toward the collapse of an entire system, a system that some consider defective but still "perfectible." Today there is no longer any such perfectible system. On the contrary, every day this system reaches new heights in all the forms of inhumanity it has been amassing for so many years.

If someone should criticize these assertions as lacking any basis, it is entirely within their rights to present a different position that is coherent. If they feel that our position is pessimistic, as humanists we affirm that the new direction toward a humanized world will prevail over this mechanical negative process. And that new direction will be propelled by the revolution that the vast communities of humanity will finally bring about, those thousands of millions of human beings who are every day denied their destiny.

With this letter I send my

Warmest regards,
Silo

Ninth Letter to My Friends

November 21, 1993

Dear Friends,

Often I receive correspondence in which people ask me what is happening today regarding human rights. I do not myself have the information necessary to provide a full answer to this question. I believe, rather, that the countries who are signatories to the Universal Declaration of Human Rights know what is happening, that is, the more than one hundred sixty nations of the world who, on December 10, 1948 or thereafter, indicated by their signatures their acceptance of this declaration, which was prepared under the auspices of the United Nations. All of them understood the issues in question, and all committed themselves to defend the rights proclaimed. Many also signed the Helsinki Accords and sent representatives to subsequent commissions on human rights and the international courts.

1. Violations of Human Rights

In reviewing the daily accounts of current events related to human rights, however, one feels compelled to reformulate this question as follows: What is this hypocritical game that governments are playing in their treatment of human rights? Even a cursory examination of the information that flows from news organizations, newspapers, magazines, radio, and television will provide an answer to this question. As one example, let us consider the Amnesty

International report for the most recent year, 1992, and briefly review some of its data.

Along with conspicuous disasters such as the wars in Yugoslavia and Somalia, violations of human rights were found to have increased all over the world. There were prisoners of conscience in 62 countries, institutionalized torture in 110, and political assassinations employed by governments in 45. The war in Bosnia-Herzegovina exhibited abuse and slaughter by all sides, perpetrated against tens of thousands of people who were assassinated, tortured, and starved, often solely because of their ethnicity. These same phenomena are also occurring in other places such as Tadzhikistan and Azerbaijan.

Accusations of torture and mistreatment by security forces increased significantly in Germany, France, Spain, Portugal, Romania, and Italy. The race of the victims in these cases was often seen to play an important role. Armed opposition groups in the United Kingdom, Spain, and Turkey also committed serious violations of human rights. In the United States 31 people were executed, the highest toll since 1977, the year that the death penalty was re-instituted. In Somalia thousands of unarmed civilians were killed during this same period.

In 1992 security forces and death squads murdered approximately 4,000 people in Latin America. In Venezuela there were dozens of arrests and executions of political prisoners during the suspension of constitutional guarantees following the attempted coups of February 4 and November 27. In Cuba approximately 300 persons were kept imprisoned for political reasons, although because international observers from Amnesty International were barred from the country the accuracy of these data could not be confirmed. In Brazil police killed 111 in São Paolo during a prison riot, while in São Paulo, Rio de Janeiro, and other areas of the country hundreds of children and other "undesirables" were murdered. In Peru 139 persons "disappeared," and security forces carried out 65 extra-judicial executions. Amnesty International also received reports of widespread abuse in Peru's rural mountain areas, and approximately 70 persons were sentenced to life imprisonment in irregular judicial proceedings. Armed opposition groups also murdered several dozen people in different regions of that country. In Colombia, repeated reports of human rights violations were denied by presidential advisors on this matter, who attributed such reports to opposition politicians seeking to distort the image of political reality in the country. Notwithstanding these denials, Amnesty International accused the armed forces and paramilitary groups of the extra-judicial executions of no less than 500 persons, while armed opposition groups and drug-trafficking organizations murdered some 200 more.

Amnesty International also reported that the struggle against militant Islamic groups triggered a deterioration of the human rights situation in various Arab countries, including Algeria and Egypt. Torture, lack of due process, political assassinations, "disappearances," and other major violations of human rights were perpetrated by government agents throughout the Middle East. In Egypt the adoption of new legislation "facilitated" torture of political detainees, and a military court sentenced to death 8 Islamic militants, presumed to be members of an armed group, following a process that was deemed unjust. In Algeria as many as 10,000 persons were interned in isolated desert concentration camps without being charged and without due process. In turn, fundamentalist groups were found to be responsible for the murder of civilians and other serious violations of human rights in Algeria and Egypt, as well as in the territories occupied by Israel. Detention without due process was particularly widespread in Syria, but also took place in Israel, Libya, Iraq, Kuwait, Saudi Arabia, Morocco, and Tunisia.

Regarding China, Amnesty International called attention to the number of "prisoners of conscience" and the sentencing of political activists without previous judicial proceedings.

News and information organizations of various leanings have prepared world maps showing dozens of countries dotted with human rights violations and other maps showing mounting death tolls from religious and inter-ethnic warfare. In addition, they highlight areas of starvation where tens of thousands of people have died, either in their homelands or during large-scale migrations.

It should be emphasized that the information outlined above does not by any means exhaust either the theme of human rights or, consequently, the forms of violation of human rights taking place in the world today.

2. Human Rights, Peace, and Humanitarianism as Pretexts for Intervention

Today there is renewed vigor in the discussion of human rights, yet the cast of those who carry this banner has changed. In decades past, progressive movements have worked actively in defense of these principles, which have been established by a consensus of the nations. Of course, even while paying lip service to these rights, many dictatorships have made a mockery of human needs and of personal and collective freedom. Some have announced that as long as citizens did not speak out against the prevailing system they would continue to have access to housing, health care, education, and employment. Logically, these governments said, we should not confuse liberty with license, and "license" is to speak out against the government.

Today it is the right wing in many countries that has raised this standard anew and tries to appear active in defense of human rights and peace, above all in those foreign countries where their own domination is not complete. Taking advantage of certain international mechanisms, they organize forces for intervention capable of reaching any point on the globe with the stated goal of imposing "peace and justice." Supporting the faction that is most subordinate to them, they begin by bringing in food and medicine, only to later attack the populace with bullets. Soon, any fifth column will be able to claim that elements in their country are disturbing the peace or that human rights are being trampled, and thus request assistance from these interventionists.

By now, primitive treaties and mutual defense pacts have been perfected into documents that legalize action by "neutral" forces. In this way, the old *Pax Romana* is being revived and introduced once more. These are, in short, ornithological avatars that, beginning with the eagle on the banner of the legionnaires, later take the form of Picasso's dove, until by the time we reach the present day we find talons growing once more beneath its bedraggled plumage. No longer does this feathered creature fly back to the biblical Ark bearing an olive branch, it now returns to the Ark of Assets with a dollar clutched in its strong beak.

Of course, all of this is well seasoned with compassionate arguments. And we should be concerned by such events, because **even when these "neutral" forces intervene in third countries for humanitarian reasons clear to all, they are setting precedents that may subsequently be used to justify new actions whose motives are neither so humanitarian nor so clear to all.** As a result of the process of globalization the United Nations is seen to be playing an increasingly military role, one that entails more than a few risks. Once again the sovereignty and self-determination of peoples are being imperiled by this manipulation of the concepts of peace and international solidarity.

Let us set aside now for another occasion themes related to peace in order to look more closely at human rights which, it is clear, are not limited solely to questions of conscience, political freedoms, and freedom of expression. Nor can protecting these rights be reduced simply to preventing the persecution, imprisonment, or deaths of citizens who have disagreements with a given government. That is, the defense of human rights cannot be limited only to defending people who are facing the actual or potential exercise of direct physical violence against them. Although certain basic ideas have been embodied in the Universal Declaration of Human Rights, there is a great deal of confusion and much uncoordinated work surrounding these issues.

3. The Other Human Rights

The second article of the Universal Declaration of Human Rights states:

> Article 2. 1. Everyone is entitled to all the rights and freedoms set forth in this Declaration, without distinction of any kind, such as race, color, sex, language, religion, political or other opinion, national or social origin, property, birth, or other status.

Among the rights enumerated are the following:

> Article 23. 1. Everyone has the right to work, to free choice of employment, to just and favorable conditions of work and also to protection against unemployment.
> Article 25. 1. Everyone has the right to a standard of living adequate for the health and well-being of himself and of his family, including food, clothing, housing and medical care, and necessary social services, and the right to security in the event of unemployment, sickness, disability, widowhood, old age, or other lack of livelihood in circumstances beyond his control.

These articles, signed by the member states, are based on the concept of the equality and universality of human rights. In neither the spirit nor the letter of the declaration do we find conditions such as: These rights will be respected *as long as they do not disturb macroeconomic variables*. Or any statement such as: The rights declared will be respected *as soon as* we become a society of prosperity. Yet the meaning of these articles could be twisted by appealing to Article 22:

> Article 22. 1. Everyone, as a member of society, has the right to social security and is entitled to realization, through national effort and international cooperation *and in accordance with the organization and resources of each State*, of the economic, social and cultural rights indispensable for his dignity and the free development of his personality. [emphasis added]

The phrase "and in accordance with the organization and resources of each state" could be used to dilute the effective exercise of these rights, and this leads us directly to the discussion of economic models.

Let us consider, for example, a country with sufficient organization and resources to enter the system of free market economies. As this transition takes place, the State will be reduced to a mere "administrator," while private enterprise will focus solely on the development of business. Budgets for health care, education, and social security will be steadily reduced as the State ceases to play a role in assisting the people. In the end, the government will no longer have obligations in these areas, nor will private enterprise assume responsibility to meet these needs. The laws that could have required business to protect these rights are being rescinded or rewritten as companies increasingly resist all regulations, even those related to the health and safety of their own employees.

But the idea and practice of privatizing health care will save the day by allowing private enterprise to fill the vacuum left during the previous stage of transition. This model will be reproduced in every field as "privatism" advances, offering its efficient services to everyone who is able to pay for them – an arrangement that will serve very well to meet the needs of some twenty percent of the population.

Who, then, will protect the universal and egalitarian conception of human rights if they are exercised "in accordance with the organization and resources of each State"? For the defenders of that ideology will continue to assure us that the smaller the State becomes, the more the economy of that country will prosper. This discussion soon passes, however, from idyllic declarations about the coming "general prosperity" to brutal statements with the character of ultimatums delivered in roughly these terms: If laws are passed that place limitations on capital, capital will flee the country and there will be no foreign investment, international loans, or refinancing of previously contracted debts. Then exports and production will fall and, in short, the whole social order will be put at risk. This displays in its stark simplicity one of the many contemporary schemes for extortion.

While the example considered above is of a country with sufficient resources to negotiate the passage toward a free market economy, it is easy to imagine how much more difficult the circumstances would be if the country in question did not possess the basic requisites of resources and organization.

As the New World Order is now proposed, and in light of economic interdependence, **in all countries, rich and poor alike, the forces of capital will try to undermine the universal and egalitarian conception of human rights.**

The previous discussion cannot be strictly derived from the grammatical terms of Article 22, because neither in that article nor elsewhere in the Universal Declaration of Human Rights is any economic consideration placed above the people that relativizes their rights. Nor is it legitimate to introduce tangential arguments by proposing, for example, that since the economy is the basis of social development, we must first dedicate all our efforts to macroeconomic variables, so that when we achieve prosperity then we will be able to attend to human rights. This is as clumsily linear as saying: Because society is subject to the law of gravity, we need to concentrate first on this problem, and only when we have solved it will we be free to speak of human rights. In a sane society no one thinks of constructing buildings on unstable foundations because everyone recognizes the conditions that gravity sets. Similarly, everyone is well aware of economic conditionings and the importance of resolving them correctly **as a function of human life.** But these digressions take us away from our theme.

The consideration of human rights cannot be reduced only to the foregoing questions of work, compensation, and assistance, just as earlier we saw they could not be limited solely to the ambits of political expression and freedom of conscience. And although there are certain defects of expression in the Universal Declaration of Human Rights, notwithstanding these it is clear that a scrupulous application of its articles by all governments would be sufficient for our world to experience a positive change of great importance.

4. The Universality of Human Rights and the Cultural Thesis

There exist diverse conceptions of the human being, and this variety of points of view is often related to the different cultures from which people observe reality. And these issues necessarily affect the question of human rights as a whole. Indeed, faced with the idea of a universal human being with the same rights and functions in all societies, today some are raising a *cultural thesis* in defense of a different position regarding these questions. The supporters of this position regard supposedly universal human rights as simply a generalization of the Western point of view in an unjustified claim of universal validity. For example, consider Article 16:

1. Men and women of full age, without any limitation due to race, nationality or religion, have the right to marry and to found a family. They are entitled to equal rights as to marriage, during marriage and at its dissolution.
2. Marriage shall be entered into only with the free and full consent of the intending spouses.
3. The family is the natural and fundamental group unit of society and is entitled to protection by society and the State.

These three sub-paragraphs in Article 16 present numerous difficulties of interpretation and application in various cultures that stretch from the eastern Mediterranean through the Middle East and into Africa and Asia – that is to say, they create difficulties for the greater part of humanity. The world is so large and so varied that over vast parts of it not even marriage and the family coincide with the parameters that seem so "natural" to the West. As a consequence, these institutions and the universal human rights associated with them are the subject of continuing debate.

The same occurs if we consider the general conceptions of law and justice. If we compare ideas regarding criminal punishment and the rehabilitation of criminals, we find no agreement on these points even among nations from the same Western cultural context. To uphold the point of view of one's own culture as valid for all of humanity, then, leads to positions that are frankly ludicrous. For example, the legal penalty of cutting off the hand of a thief as practiced in certain Arab countries is viewed as a clear violation of human rights in the United States – while at the same time they like to hold academic debates on whether to execute criminals by the use of cyanide gas, 2,000 volts of electricity, lethal injection, hanging, or some other macabre delight of capital punishment. It should be noted, however, that just as in the United States a significant proportion of the society rejects capital punishment, so too in Arab countries many oppose corporal punishment for those who have broken the law.

Even the West itself, swept along by changing practices and customs, is having great difficulty in trying to uphold its traditional idea of the "natural" family. Can a family today contain adopted children? Of course it can. Can a family have spouses who are members of the same sex? Some legislatures already allow this. What, then, defines the family – its "natural" character or the voluntary commitment of people to fulfill certain functions? On what basis can we say that the monogamous family of some cultures is better than the polygamous one of others? And if this is the state of the discussion, can we continue to speak of a single set of laws that is universally applicable to the family? Which human rights are to be defended – and which are not – regarding the institution of the family?

Clearly, the dialectic between the universalist thesis (hardly universal even in its own culture) and the cultural thesis cannot be resolved in the case of the family (which I have considered as only one of many possible examples), just as I am afraid that for now it will remain similarly unresolved for other areas of the social endeavor.

To sum this up: Here we find in play a general conception of the human being that is not sufficiently well-founded to encompass the many positions in conflict. Yet the need for such a comprehensive conception is evident, because neither the law in general nor human rights in particular can prevail if their deepest meaning is not clear.

No longer can we raise the most general questions of law only in the abstract. **Either we are dealing with rights that, to have effect, must flow from established power, or we are dealing with rights that are only aspirations yet to be fulfilled.** In regard to the issue of rights, I have written elsewhere (see the chapter "Law" in *The Human Landscape*[6]):

> Practical people who have not become lost in theorizing have declared that law is necessary in order for there to be social coexistence. It is also said that the law is made to defend the interests of those who impose it.
>
> It seems that in the situation previous to power a particular law is installed, which in turn legitimizes that power. So it is that power, as the imposition of an intention – whether accepted or not – is the central issue. It is said that force does not generate rights, but paradoxically this statement is normally accepted only when force is thought of as brutal physical fact, when in reality force – economic, political, and so on – does not need to be expressed perceptually to make its presence felt and to demand respect. In any case, even physical force, that of arms, for example, expressed as naked threat creates situations that are justified legally, and we cannot deny that the use of arms in one direction or another depends on human intention and not on a right.

And further on:

> All those who violate the law are ignoring a situation that is asserted in the present, exposing their temporality – their future – to the decisions of others. But it is clear that this "present" in which the law begins to take effect has its roots in the past. Customs, morality, religion, or social consensus are the sources

customarily invoked to justify the existence of the law. Each depends in turn on the power that imposes it. And these sources are changed when the power that gave them origin declines or transforms so that maintaining the previous judicial order begins to clash with what is "reasonable," with "common sense," and so on. When the legislature repeals or rewrites a law, or a group of representatives of the people amend a country's basic charter or constitution, they apparently do so without violating the law in general, because they are not subject to the decisions of others, because they hold power or act as the representatives of established power, and in this situation it is clear that power generates rights and obligations and not the reverse.

To end, let me cite the following:

Human rights do not have the universal application that would be desirable because they do not flow from the universal power of the human being, but only from the power that one part now exercises over the whole. If even the most elementary claims to the governing of one's own body are trampled underfoot in all latitudes, then we can speak only of aspirations yet to become rights. **Human rights do not pertain to the past, they lie ahead in the future, calling our intentionality, sustaining a struggle that is rekindled in each new violation of humanity's destiny. For this reason, every protest in favor of human rights has meaning because it shows the powers that be that they are not omnipotent and that they do not control the future.**

As for our general conception of the human being, it does not seem necessary to review it here or to reaffirm that the recognition we give to diverse cultural realities does not invalidate the existence of a common human structure that is in historical flux in a converging direction. The struggle to establish a *universal human nation* is also the struggle, from each culture, to put into practice human rights that are ever more coherently defined.

If the right to a fulfilled life and freedom is suddenly ignored in a certain culture and other values placed above the human being, it is because something there has gone astray, something is diverging from our common destiny. Should this happen, then the expression of that culture *in that precise point* must be clearly repudiated.

It is true that the formulations of human rights in the Universal Declaration of Human Rights are imperfect, but for now this is all that we have at hand to defend and to perfect. **Today these rights are still considered aspirations that cannot be fully realized given the established powers. The struggle for the full application of human rights leads necessarily to questioning the powers that be, orienting action toward replacing them with the powers of a new and human society.**

With this letter I send my

Warmest regards,
Silo

Tenth Letter to My Friends

December 15, 1993

Dear Friends,

Toward what destiny are present-day events heading? Optimists feel that we will soon find ourselves in a worldwide society of abundance in which society's problems will be solved – a sort of paradise on Earth. Pessimists believe that current symptoms indicate a growing sickness of both institutions and human groups – the entire population and ecological system – a sort of hell on Earth. In contrast, those who view historical mechanisms as relative feel that everything rests on our present behavior – that heaven or hell depend on our actions. Of course, there are others not in the least interested in what happens to anyone other than themselves.

Among these varied opinions, the important one to us is that the future depends on what we do today. Yet even within this position there are differences of approach.

Some say that, since this crisis has been brought on by the voracity of the banking system and the multinational companies, when these problems reach the point of endangering their interests they will set mechanisms of recovery in motion, just as they have done on previous occasions. In regard to action, such people favor gradually adapting to the reform processes they claim are converting capitalism to the benefit of the majorities.

Others argue that we cannot let everything depend solely on the good will of the few, and what is required therefore is to demonstrate the will of the majority through political action and by educating the people, who now live in a situation of extortion under the dominant scheme of things. According to them, a moment of general crisis for the system will come, and it will be important to take advantage of this for the cause of the revolution.

Finally, there are those who maintain that capital as well as labor, all cultures, nations, and organizational forms, all artistic and religious expressions, all human groups and individuals are caught up in a process of technological acceleration and destructuring that is beyond their control. Flowing out of a long historical process, things today have reached a point of worldwide crisis that is affecting every political and economic scheme. And both the general process of disorganization and the general recovery will proceed independently of any such schemes.

Those who uphold this structural point of view stress that it is necessary to forge a global understanding of these phenomena at the same time that one acts locally in societal, group, and personal areas of some minimum specificity. Given how interconnected the world is, they do not believe that any step-by-step gradualism society will supposedly adopt over time can be successful – instead they strive to generate a series of *demonstration effects* sufficiently energetic to produce a general inflection in the process.

They therefore champion the constructive capacity of human beings to unite and transform economic relationships, to change institutions, and to struggle tirelessly in dismantling all the factors that are bringing about a regressive involution with no way out. As contemporary humanists we hold this last position. Clearly, of course, this as well as the previous descriptions have been simplified, omitting the multiplicity of variants that can be derived from each of them.

1. Destructuring and Its Limits

It is pertinent here to point out the limits of political destructuring, which will not stop until this process reaches down to the base of society and every individual.

Let us consider some examples. The weakening of centralized political power is more evident in some countries than in others. As they take advantage of the growing strength of autonomous regions or the pressure of secessionist movements, certain interest groups or simple opportunists wish to stop the process of destructuring at exactly the point that will leave control of the situation in their hands. According to their aspirations, once a canton has seceded, a new republic has separated from the former nation, or an autonomous region has been freed from the central power, it ought now to continue as the new organizational structure.

What happens instead is that these new powers are in turn challenged by the micro-regions, counties, cities, and towns that lie within them. None of these constituent units can see why an autonomous region that has been freed from a former central power should now centralize power over its component areas, no matter how vigorously the new region may offer as rationales the sharing of language, a common folklore, or even some ineffable "historical and cultural collectivity." This is because when it comes to paying taxes and allocating budgets, the relevance of folklore extends only as far as tourism and record companies. And were the cities to be freed from the newly independent region, the neighborhoods would apply this same logic, and so on down the scale until this reaches even the neighbors who live on opposite sides of the street.

Then someone may say, "Why should those of us who live on this side of the street have to pay the same taxes as those on the other side? We have a higher standard of living, and our taxes are only going to solve the problems of those other people who don't even try to get ahead through their own efforts. It's better for each to take care of their own." And so on down the scale until one hears the same concerns expressed even in the individual houses in the neighborhood – and no one will be able to stop this mechanical process at precisely the stage that interests them. That is, things will not come to a stop in a simple process of medieval-style feudalization, a situation that corresponded to small, thinly scattered populations whose sporadic contact and interchange took place through means of communication controlled by quarreling feudal lords or bands of toll and tax collectors. Today's situation does not at all resemble that of previous eras in terms of production, consumption, technology, communications, population density, and many other factors.

At the same time, economic blocs and common markets will increasingly absorb the decision-making power that nations formerly held. In a given area, newly autonomous regions will be able to escape from their former national entity, but at the same time cities or groups of cities within them will bypass the old administrative levels, seeking inclusion as full members in the new regional superstructure. And the regional economic entities will give serious consideration to those independent regions, cities, or groups of cities that possess strong economic potential.

In the economic warfare among the various regional blocs, there is nothing to prevent certain member countries from beginning to establish "bilateral" or "multilateral" relations with other areas, thus escaping the orbit of the regional market of which they form a part. Why couldn't the United Kingdom, for example, establish closer ties with the NAFTA, beginning at first with a few exceptions to existing European arrangements. Later on, depending on the progress of the relationship, what would stop it from eventually abandoning its former market to join the North American regional market? Or if Quebec were to secede from Canada, what would keep it from opening negotiations outside the region of the NAFTA? In Latin America it is clear that organizations such as the Latin American Free Trade Association (ALALC) or the Andean Pact (Pacto Andino) are no longer viable, as already we see Columbia and Chile beginning to integrate their economies with an eye to inclusion in the NAFTA, even as the Southern Common Market (MERCOSUR) is affected by possible regional secessions within Brazil.

Moreover, if Turkey, Algeria, and countries of the northern Mediterranean begin to join the European Common Market, other countries that are excluded could tend to strengthen their mutual ties and negotiate as a group with regional markets of other geographical areas. And while powers such as China and Russia as well as the countries of Eastern Europe continue to undergo rapid centrifugal transformations, what effects will this have on the regional blocs as they are now visualized?

While it is unlikely that things will turn out exactly as described in these examples, the tendency toward regionalization may well take unexpected turns, resulting in arrangements quite different from those schemes now proposed based on geographical contiguity, and therefore relying on conventional geopolitical prejudices. So it is that fresh disturbances may befall today's newly laid schemes and strategies, whose objectives go beyond simple economic union and include the intention to form political and military blocs.

Since in the end it will be the forces of big capital that decide things based on what is most favorable for the evolution of their businesses, no one should imagine with too much certainty regional maps drawn as in the past in accordance with geographical contiguity in which highway and rail links radiating from central points play the principal role. The trend today is toward arrangements redesigned around high-volume air and ship traffic supported by worldwide satellite communications.

Even by colonial times, geographical proximity had already been replaced by the far-flung overseas checkerboard of the great powers, which with the two world wars entered decline. For some, the present rearrangements take the problem back to pre-colonial stages, and they imagine that an economic bloc must be organized in a spatial continuum, through which they project their own particular nationalism into a sort of regional "nationalism."

In short, the limits of destructuring are not given in particular by those countries or autonomous regions newly freed from a central power or in general by economic regions organized according to geographical contiguity. The lower limits of destructuring reach right down to each neighbor and individual, while the upper limits reach the world community as a whole.

2. Some Important Areas of the Phenomenon of Destructuring

Among many possible areas in the process of destructuring, I would like to focus on three areas in particular: the political, the religious, and the generational.

It is clear that in general various political parties, arising from time to time as "right," "center," and "left," will alternate holding the now-reduced power of the State. Already we are seeing many "surprises," and still others are in store as forces long supposed to have disappeared emerge once more, and coalitions and alignments enthroned for decades dissolve amid widespread scandal. While this is nothing new in the game of politics, what is genuinely original is that ostensibly opposed political factions are succeeding each other without altering in the slightest the process of destructuring, which of course affects them too. And in regard to the proposals, language, and style of politics, we will witness a general syncretism in which ideological profiles fuse, growing more blurred with each passing day.

Faced with this battle of slogans and empty forms, average citizens will continue to distance themselves from any kind of participation, to concentrate only on what is most immediate and perceptual. But social discontent will continue to intensify, making itself felt through spontaneous protests, civil disobedience, outbursts of unrest, and the appearance of psycho-social phenomena with explosive growth. In these circumstances, new forms of irrationalism are emerging and, with various forms of intolerance as their rallying cries, growing dangerously close to gaining ascendancy.

In light of this, it is clear that if a central power wishes to stifle demands for independence, it will feel moved to adopt increasingly radical positions in order to draw other political groups into its sphere. What party will be able to remain uninvolved – at the risk of losing its influence – if violence sparked by territorial, ethnic, religious, or cultural disputes explodes in a given point?

Political factions will have to take positions on such issues, as we see today in various parts of Africa (where there are 18 points in conflict); the Americas (4 points in conflict – Brazil, Canada, Guatemala, and Nicaragua, without including the claims of indigenous peoples in Ecuador and other countries of the Americas, or the deteriorating racial situation in the United States); Asia (10 points in conflict, counting the Chinese-Tibetan conflict, but without considering the inter-canton differences arising throughout China); south and Pacific Asia (12 points, including the protests of the indigenous peoples of Australia); Western Europe (16 points); Eastern Europe (4 points, counting the Czech Republic and Slovakia, the former Yugoslavia, Cyprus, and the former Soviet Union as only one point each; there are over 30 points in conflict if we include the many countries of the Balkans and the former Soviet Union, which has inter-ethnic and border problems in more than 20 republics stretching well beyond Eastern Europe); and the eastern Mediterranean and the Middle East (9 points in conflict).

Politicians will also be moved to echo the increasing radicalization that the traditional religions are experiencing, such as that between Muslims and Hindus in India and Pakistan, between Muslims and Christians in the former Yugoslavia and Lebanon, and between Hindus and Buddhists in Sri Lanka. They will have to respond to the inter-sect fighting within a given religion such as that between Sunnis and Shiites in the sphere of influence of Islam, and between Catholics and Protestants in the sphere of influence of Christianity. They will be drawn to participate in the religious persecution that has begun, first in the West, through the press and the passing of laws restricting freedom of religion and conscience.

It is clear that the traditional religions will try to impede the newer religious forms that are now awakening all over the world. According to the "experts" and pundits, who are normally atheists but objectively allied with the dominant sect of their area, the harassment of the new religious groups "does not constitute a limitation on freedom of thought, but rather a protection for the freedom of belief that now finds itself under attack by the brainwashing of the new cults, which, furthermore, are undermining our civilization's traditional values, culture, and way of life."

In this way, politicians usually far removed from the theme of religion are beginning to take part in this witch-hunt because, among other things, they note the massive popularity that these new expressions of faith – which also carry an undercurrent of revolution – are beginning to achieve. No longer will they be able to claim as in the nineteenth century that "religion is the opiate of the masses." No longer can they speak of the slumbering isolation of the masses and the individual, when Muslim populations are proclaiming the establishment of Islamic republics, when in Japan (with the collapse of the national religion of Shintoism following World War II) Buddhism formed the motor that carried the Komeito to power, when the Catholic Church is launching new political ventures in the wake of the exhaustion of Christian Socialism and Third Worldism in Latin America and Africa. In any event, the atheist philosophers of the new times will have to change the terms of their discourse, replacing the phrase "opiate of the masses" with the phrase "amphetamine of the masses."

Leaders will also have to take positions regarding youth, increasingly characterized as constituting a "threat to society," with dangerous tendencies toward drugs, violence, and lack of communication. Those leaders who persist in ignoring the profound roots of these problems will be in no position to give satisfactory answers simply by inviting young people to participate in conventional politics or the traditional cults, or to enjoy the offerings of a decadent civilization controlled by money. Meanwhile, such leaders are contributing to the psychic destruction of an entire generation and the rise of despicable new economic powers that grow rich by preying on the anguish and psychological alienation of millions of human beings.

Many leaders now ask in surprise where this growing violence among young people is coming from – as if it were not these leaders themselves, the former or current generations to hold power, who have overseen the perfecting of a systematic violence, exploiting even the advances in science and technology to make their manipulations ever more efficient.

Some point to a supposed "autism" among youth and, based on this view, attempt to establish relationships between the increasing life spans of adults and the longer period of education and training required before young people are allowed to enter full participation. This explanation, while not without basis, is certainly not sufficient to understand these more ample processes. What we can observe is that the generational dialectic, the motor of history, has become temporarily stalled, and with this a dangerous abyss has opened between two worlds.

Here it is interesting to recall that over two decades ago, when a certain thinker warned of these incipient tendencies that today we find expressed in substantive problems, those fine Mandarins, flanked by their "experts" and formers of opinion, succeeded only in tearing their vestments in frenzied accusations that it was just such discourse that was, in addressing these problems, somehow *causing* the war between the generations.

In those times, a powerful force of youth that should have heralded the advent of a new phenomenon as well as the creative extension of the historical process, was diverted by the diffuse exigencies of the decade of the sixties, and pushed into a dead-end guerrilla struggle in various parts of the world.

Further problems are sure to befall those who now expect the new generations simply to channel all their desperation into tumultuous music or the sports stadium, limiting their protests to t-shirts and posters bearing innocent slogans. The situation of asphyxia for young people creates irrational and cathartic conditions that are ripe to be channeled by fascists, authoritarians, and the violent of all types. Nor is sowing seeds of mistrust and viewing every young person as a potential criminal the way to reestablish a dialogue between the generations. No one, moreover, is showing any enthusiasm for allowing the new generations access to society's communications media, nor are those in control inclined toward public discussion of these issues unless they are dealing with "model youth" who, accompanied by rock music, simply parrot the established political wisdom or venture forth in the spirit of Boy Scouts to clean seabirds covered with oil – but without questioning the forces of big capital, which continue to produce these ever-widening ecological disasters!

I fear that any genuine youth organization (whether student, artistic, labor, or religious) will be suspected of the worst kinds of misdeeds simply because they are not sponsored by a union, political party, foundation, or church. Despite so much manipulation of young people, there are still those who ask why youth do not embrace the marvelous proposals proffered by the established powers, adding that it would be to the benefit of these future citizens to busy themselves with study, work, and sports. Were this to occur, no one would have to worry about any "lack of responsibility" among such busy young people.

However, if unemployment should continue to climb, if the recession should become chronic, if everywhere the phenomenon of marginalizing and neglecting young people should continue to grow, then we shall see what today's lack of participation develops into. For various reasons – wars, hunger, unemployment, moral fatigue – the generational dialectic itself has become destructured, producing a silence that has lasted for two long decades, a silence now being shattered by heart-rending cries and acts of desperation that lead nowhere.

In light of all of the above, it seems abundantly clear that no one will be able to reasonably orient the processes of a world that is fast dissolving. While this dissolution is tragic, it is at the same time illuminating the birth of a new civilization – the world civilization. And if this is happening, then a certain type of collective mentality must also be disintegrating, as a new way of being conscious of the world emerges. Regarding this point, I would like to include here something said in the first letter:

A new sensibility is being born that corresponds to these changing times. It is a sensibility that grasps the world as a whole – an awareness that the problems people experience in one place involve other people, even if they are far away. Increasing communication, trade, and the rapid movement of entire human groups from one place on the planet to another all attest to this growing process of globalization.

As the global character of more and more problems comes to be understood, new criteria for action arise. There is an awareness that the work of those who desire a better world will be effective only if they make their efforts grow outward from the environment where they already have some influence. In sharp contrast to other times, so full of empty phrases meant only to garner external recognition, today people are beginning to find value in humble and deeply felt work, work done not to enhance one's self-image, but rather to change oneself and bring about change in one's immediate environment of family, work, and friendship.

Those who truly care for people do not disdain this work done without fanfare, this work that proves so incomprehensible to those opportunists who were formed in an earlier landscape of leaders and masses – a landscape in which they learned well how to use others to catapult themselves to society's heights.

When a person comes to the realization that schizophrenic individualism is a dead end, when they openly communicate what they are thinking and what they are doing to everyone they know without the ridiculous fear of not being understood, when they approach others not as some anonymous mass but with a real interest in each person, when they encourage teamwork in both the interchange of ideas and the realization of common projects, when they clearly demonstrate the need to spread this task of rebuilding the social fabric that others have destroyed, when they feel that even the most "unimportant" person is of greater human quality than some heartless individual whom circumstance has elevated to what is, for now, the pinnacle of success – when all this happens it is because within this person destiny has once again begun to speak, the destiny that has moved entire peoples along their best evolutionary path, the destiny that has been so many times distorted and so many times forgotten, but is always reencountered in the twists and turns of history.

Today we can glimpse not only a new sensibility and a new mode of action but also a new moral attitude and a new tactical approach to facing life.

Today, hundreds of thousands of people all over the world affirm the ideas embodied in the "Humanist Document" (see Sixth Letter, p. 69). They are Communist-Humanists, Socialist-Humanists, Liberal-Humanists, Environmentalist-Humanists, and a great many others, all of whom without abandoning their own causes take one step toward the future. They are people who struggle for peace, for human rights, and for an end to discrimination. Among them are, of course, both atheists and those who have faith in human beings and their transcendence. And all of them have in common a passion for social justice, an ideal of human brotherhood based on the convergence of diversity, a disposition to leap beyond all prejudice, and a coherent personality in which their personal lives are not separate from the struggle for a new world.

3. Targeted Action

There are still political militants who worry about who will be the next president, prime minister, senator, or representative. It is possible that they do not yet fully comprehend the real extent of the destructuring toward which everything is heading and how little any of these "hierarchs" will mean for the transformation of society. There will also be more than one case in which such anxiety is linked to the personal situation of these supposed militants, who are worried about their own position in the world of political deal-making.

The key question in any case is for people to focus on **understanding how to establish priorities among the conflicts in the places where they carry out their daily lives and to know how to organize valid and effective fronts of action based on such conflicts.**

In each situation it is important to understand what characteristics are required to form grassroots committees on health, education, labor, student, and other issues, and what characteristics are necessary for *centers for direct communication* and networks of *neighborhood councils*. It needs to be clear how to give participation to even the smallest and least noticed of those organizations through which people express their work, culture, sports, and religiosity.

Here it is useful to explain that when we refer to people's immediate environment of coworkers, family, and friends, we are emphasizing in particular the *places* in which these relationships occur.

Speaking in spatial terms, the minimum unit of action is the *neighborhood*, for it is here that people feel each conflict, even though the roots of that conflict may be far away. A *center for direct communication* forms a place in the neighborhood where people can directly discuss all economic and social problems, as well as all the problems of health care, education, and the quality of life in general.

The political focus is to give a higher priority to the neighborhoods than to the city, county, state, province, or even a newly independent region or the country as a whole. In truth, long before nations were formed, people congregated together in human communities where, as they put down roots, they became neighbors. Later on, administrative superstructures were set up that increasingly robbed the neighborhoods of their autonomy and power. Yet the legitimacy of any given order derives only from the inhabitants, from these neighbors, and it is from them that all representation in a real democracy must arise.

Every town and city should be in the hands of its neighborhoods and, if this is the case, no one can coherently propose the objective of setting up multiple layers of representatives or deputies, as occurs in leader-dominated hierarchical politics. Rather, all such arrangements can only be the *result* of the grassroots operation of the organized social base. The concept of *neighborhood* applies to populations that are spread out, as well as to those concentrated within a limited area or living in large apartment buildings or complexes.

It is important for the neighborhoods to decide among themselves, through the structures that connect them, the status of their district. And their decisions within this district should, of course, not depend on some faraway superstructure that simply dictates orders.

When several neighborhoods set a *humanist district action plan* in motion, and their district, town, or city proceeds to organize real democracy, this *demonstration effect* will make itself felt far beyond the boundaries of that bastion. And rather than proposing a gradualism through which this new approach will little by little gain territory until finally it has spread to every corner of a country, what is key is to demonstrate in practice that in at least one place a new system is working.

The detailed problems presented by all of the above are of course numerous, and it would be beyond the scope of this letter to attempt to treat them here.

With this final letter I send my

Warmest regards,
Silo

Notes

Fourth Letter to My Friends

1. Silo, *Contributions to Thought*, forthcoming from Latitude Press. [*Silo: Obras Completas, Volumen I, Contribuciones al pensamiento*, (San Diego: Latitude Press, 1993), 141-200.]
2. Silo, *Humanize the Earth, a Trilogy: The Inner Look, The Inner Landscape, The Human Landscape*, forthcoming from Latitude Press [*Silo: Obras Completas, Volumen I, Humanizar la tierra, El paisaje humano*, 53 – 80.]

Seventh Letter to My Friends

3. G. Petrovich, "La necesidad de un concepto de revolución," *La filosofía y las ciencias sociales, Primer coloquio nacional de filosofía*, Morelia, Mexico, 4–9 de agosto de 1975, (Editorial Grijalbo, 1976). ["The Need for a Concept of Revolution," *Philosophy and Social Sciences,* First National Colloquium on Philosophy, Morelia, Mexico, August 4 – 9, 1975.]
4. Silo, *Contributions to Thought*, Historiological Discussions, chapter III, sections 2 and 3. [*Silo: Obras Completas, Volumen I, Contribuciones al pensamiento*, Discusiones historiológicas, cap. III, par. 2 y 3, pp. 184 – 190.]

Eighth Letter to My Friends

5. Silo, "La Necesidad de una Posición Humanista en las Fuerzas Armadas Contemporáneas," Conferencia internacional sobre Humanización de las actividades militares y reforma de las Fuerzas Armadas, patrocinada por el Ministerio de Defensa de la CEEII, Moscú, 24–28 de mayo de 1993. ["The Need for a Humanist Position in the Contemporary Armed Forces," International Conference on Humanizing Military Activities and Reform of the Armed Forces, sponsored by the Ministry of Defense of the Commonwealth of Independent States, Moscow, May 24 – 28, 1993.]

Ninth Letter to My Friends

6. Silo, *Humanize the Earth.* [*Silo: Obras Completas, Volumen I, Humanizar la tierra, El Paisaje humano, La Ley, 65-66.*]

New Humanism Series

The New Humanism Series consists of contemporary fiction and nonfiction addressing the growing interest in humanizing individual and social life in today's world.

Letters to My Friends by Silo
Tales for Heart and Mind by Silo (also in audio)
Self Liberation by L.A. Ammann
Historical Interpretations of Humanism by Salvatore Puledda
Humanize the Earth by Silo
Contributions to Thought by Silo
Universal Root Myths by Silo
Day of the Winged Lion and Other Stories by Silo
Silo: Obras Completas, Volumen I (in Spanish)
The Tokarev Report by Salvatore Puledda

New Humanism Series books are available at fine bookstores or may be ordered direct from the publisher. Please contact us for a Free Catalog or information on quantity discounts.

Latitude Press
P.O. Box 231516
Encinitas, CA 92023-1516

800-LATITUDE (800-528-4883) • FAX 619-632-6359
Internet 73250.574@compuserve.com